SCIENTIFIC RESEARCH IN POISONOUS PLANTS

SCIENTIFIC RESEARCH IN POISONOUS PLANTS

Dr. Manoj Kumar Singh

RANDOM PUBLICATIONS

NEW DELHI - 110 002 (INDIA)

Scientific Research in Poisonous Plants

ISBN 978-93-51116-84-4

Published in 2015 in India by

Reprint 2021

RANDOM PUBLICATIONS

4376-A/4B, Gali Murari Lal, Ansari Road
New Delhi-110 002
Phone: +9111-43580356, 23289044
E-mail: randomexports@gmail.com; sales@randompublications.com; info@randompublications.com

Type Setting by: Friends Media, Delhi-110089
Printed at : Mehra Printers, Delhi-110092

Preface

We eat many plants, herbs and so forth in our daily diet. But, we must remember to be choosy. Some plants, trees or shrubs are potential killers of man. Some part of the ornamental plants or flowers in your yard may contain deadly poison. Many poisonous plants are so common and seemingly innocuous you do not suspect their toxic qualities. Many native and exotic plants are poisonous to humans when ingested or if there is skin contact with plant chemicals. However, the most common problems with poisonous plants arise from contact with the sap oil of several ever-present native plants that cause an allergic skin reaction—poison ivy, poison oak, and poison sumac. Poison ivy, poison oak, and poison sumac release an oil, urushiol, when the leaf or other plant parts are bruised, damaged, or burned. When the oil gets on the skin an allergic reaction, referred to as contact dermatitis, occurs in most exposed people as an itchy red rash with bumps or blisters. When exposed to 50 micrograms of urushiol, an amount that is less than one grain of table salt, 80 to 90 percent of adults will develop a rash. The rash, depending upon where it occurs and how broadly it is spread, may significantly impede or prevent a person from working. Although over-the-counter topical medications may relieve symptoms for most people, immediate medical attention may be required for severe reactions, particularly when exposed to the smoke from burning these poisonous plants. Burning these poisonous plants can be very dangerous because the allergens can be inhaled, causing lung irritation.

Plants cannot move to escape their predators, so they must have other means of protecting themselves from herbivorous animals. Some plants have physical defences such as thorns, spines and prickles, but by far the most common protection is chemical. Over millennia, natural selection has produced a complicated and vast array of chemical compounds that deter herbivores. Tannin is a compound that emerged relatively early in the evolutionary history of plants, while more complex molecules such as polyacetylenes are found in younger groups of plants such as the Asterales. Many of the plant defence compounds arose to defend against consumption by insects, although when livestock or humans consume such plants, they may also experience negative effects, ranging from mild discomfort to death. The old saying "Leaves of three, Let it be!" is a helpful reminder for identifying poison ivy and oak, but not poison sumac which usually has clusters of 7-13 leaves. Even poison ivy and poison oak may have more than three leaves and their form may vary greatly depending upon the exact species encountered, the local environment, and the season. Being able to identify local varieties of these poisonous plants throughout the seasons and differentiating them from common nonpoisonous look-a-likes are the major keys to avoiding exposure. More than 700 plants have been identified as producing physiologically active or toxic substances in sufficient amounts to cause harmful effects in animals. Poisonous plants produce a variety of toxic substances and cause reactions ranging from mild nausea to death. Certain animal species may have a peculiar vulnerability to a potentially poisonous plant. Xylitol is a sugar alcohol — an artificial sweetener created from birch, raspberries, plums and corn. This sweetener is found in many human "sugar free" products, such as gum, candies and other sweets. In humans, high doses may have a mild laxative effect, but in dogs, ingestion could be fatal.

The present book is essential reading for students of biology, toxicologists concerned with animal and human health, food industry regulators and plant scientists.

I thank all members of my team who have helped in the preparation of the book. My special thanks go to "Random Publications" who have published the book.

— Dr. Manoj Kumar Singh

Contents

Chapter 1

Introduction

Apocynum Cannabinum

Apocynum cannabinum (Dogbane, Amy Root, Hemp Dogbane, Indian Hemp, Rheumatism Root, or Wild Cotton) is a perennial herbaceous plant that grows throughout much of North America - in the southern half of Canada and throughout the United States. It is a poisonous plant: *Apocynum* means "poisonous to dogs".

All parts of the plant are poisonous and can cause cardiac arrest if ingested. The *cannabinum* in the scientific name and the common names Hemp Dogbane and Indian Hemp refer to its similarity to *Cannabis* as a fibre plant, rather than as a source of a psychoactive drug

Although dogbane is poisonous to livestock, it likely got its name from its resemblance to a European species of the same name.

Distribution and Habitat

Apocynum cannabinum grows in open wooded areas, ditches, and hillsides, and prefers moist places.

Description

Apocynum cannabinum grows up to 2 metres/6 feet tall. The stems are reddish and contain a milky latex capable of causing skin blisters. The leaves are opposite, simple broad lanceolate, 7–15 cm long and 3–5 cm broad, entire, and smooth on top with white hairs on the underside.

The flowers are produced in mid summer, with large sepals, and a five-lobed white corolla.

Invasive Species

In gardens it can be invasive, growing from spreading roots. When growing among corn, *Apocynum cannabinum* can reduce yields by up to 10% and when growing among soybeans, by up to 40%. It can be controlled through mechanical means, although it is difficult to control with herbicides.

Uses

Phytoremediation: *Apocynum cannabinum* is a phytoremediation plant, a hyperaccumulator used to sequester lead in its biomass.

Fibre

Apocynum cannabinum was used as a source of fibre by Native Americans, to make hunting nets, fishing lines, clothing, and twine. It is called *qéemu* [qýĐmu] in Nez Perce and [tax •hìs] in Sahaptin.

Medicinal

It is also used in herbal medicine to treat syphilis, rheumatism, intestinal worms, fever, asthma, and dysentery. Although the toxins from the plant can cause nausea and catharsis, it has also been used for slowing the pulse, and it is also a sedative and mild hypnotic.

The Concow tribe call the plant pö (Konkow language).

Arctotheca Calendula

Arctotheca calendula is a plant commonly known as cape weed, cape dandelion, or cape marigold because it originates from the Cape Province in South Africa. It is listed as a noxious weed in California and is an invasive weed in Australia as well.

The plant is a squat perennial or annual which grows in rosettes and sends out stolons and can spread across the ground quickly. The leaves are covered with white woolly hairs, especially on their undersides. The leaves are lobed or deeply toothed. Hairy stems bear daisy-like flowers with small yellow petals that sometimes have a green or purple tint surrounded by white or yellow ray petals extending further out from the flower centres.

It is cultivated as an attractive ornamental groundcover but has invasive potential when introduced to a new area. The plant can reproduce vegetatively or via seed. Seed-bearing plants are most likely to become weedy, taking hold most easily in bare or sparsely vegetated soil or disturbed areas.

Horseradish

Horseradish (*Armoracia rusticana*, syn. *Cochlearia armoracia*) is a perennial plant of the Brassicaceae family (which also includes mustard, wasabi, broccoli, and cabbage).

The plant is probably native to southeastern Europe and western Asia. It is now popular around the world. It grows up to 1.5 metres (5 feet) tall, and is mainly cultivated for its large, white, tapered root.

The intact horseradish root has hardly any aroma. When cut or grated, however, enzymes from the now-broken plant cells break down sinigrin (a glucosinolate) to produce allyl isothiocyanate (mustard oil), which irritates the mucous membranes of the sinuses and eyes. Once exposed to air (via grating) or heat, if not used immediately or mixed in vinegar, the grated mash darkens, loses its pungency, and becomes unpleasantly bitter-tasting.

History

Horseradish has been cultivated since antiquity. According to Greek mythology, the Delphic Oracle told Apollo that the horseradish was worth its weight in gold. Horseradish was known in Egypt in 1500 BC. Dioscorides listed horseradish under *Thlaspi* or *Persicon*; Cato discusses the plant in his treatises on agriculture, and a mural in Pompeii shows the plant.

Horseradish is probably the plant mentioned by Pliny the Elder in his *Natural History* under the name of *Amoracia*, and recommended by him for its medicinal qualities, and possibly the Wild Radish, or *raphanos agrios* of the Greeks. The early Renaissance herbalists Pietro Andrea Mattioli and John Gerard showed it under *Raphanus*.

Both root and leaves were used as a medicine during the Middle Ages and the root was used as a condiment on meats in Germany, Scandinavia, and Britain. It was brought to North America during Colonial times.

William Turner mentions horseradish as *Red Cole* in his "Herbal" (1551–1568), but not as a condiment. In *The Herball, or Generall Historie of Plantes* (1597), John Gerard describes it under the name of *raphanus rusticanus*, stating that it occurs wild in several parts of England. After referring to its medicinal uses, he says:

"[T]he Horse Radish stamped with a little vinegar put thereto, is commonly used among the Germans for sauce to eat fish with and such like meates as we do mustarde."

The word *horseradish* is attested in English from the 1590's. It combines the word *horse* (formerly used as an adjective meaning "strong, large, or coarse") and the word *radish*. Despite the name, this plant is poisonous to horses.

Cultivation

Horseradish is perennial in hardiness zones 2–9 and can be grown as an annual in other zones, although not as successfully as in zones with both a long growing season and winter temperatures cold enough to ensure plant dormancy. After the first frost in the autumn kills the leaves, the root is dug and divided.

The main root is harvested and one or more large offshoots of the main root are replanted to produce next year's crop. Horseradish left undisturbed in the garden spreads via underground shoots and can become invasive. Older roots left in the ground become woody, after which they are no longer culinarily useful, although older plants can be dug and re-divided to start new plants.

Pests and Diseases

Widely introduced by accident, "cabbageworms", the larvae of *Pieris rapae*, the Small White Butterfly, are a common caterpillar pest in horseradish. The adults are white butterflies with black spots on the forewings that are commonly seen flying around plants during the day.

The caterpillars are velvety green with faint yellow stripes running lengthwise down the back and sides. Full grown caterpillars are about 1-inch (25 mm) in length. They move sluggishly when prodded. They overwinter in green pupal cases.

Adults start appearing in gardens after the last frost and are a problem through the remainder of the growing season. There are three to five overlapping generations a year. Mature caterpillars chew large, ragged holes in the leaves leaving the large veins intact. Handpicking is an effective control strategy.

Culinary Uses

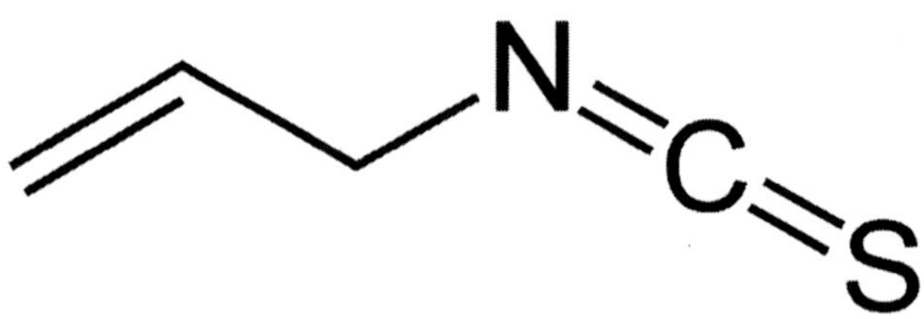

Figure: *Allyl isothiocyanate is the pungent ingredient in fresh horseradish sauce.*

Cooks use the terms "horseradish" or "prepared horseradish" to refer to the grated root of the horseradish plant mixed with vinegar. Prepared horseradish is white to creamy-beige in colour. It will keep for months refrigerated but eventually will darken, indicating it is losing flavour and should be replaced. The leaves of the plant, while edible, are not commonly eaten, and are referred to as "horseradish greens".

Horseradish Sauce

Horseradish sauce made from grated horseradish root and vinegar is a popular condiment in the United Kingdom and in Poland. In the UK it is usually served with roast beef, often as part of a traditional Sunday roast, but can be used in a number of other dishes also, including sandwiches or salads. A variation of horseradish sauce, which in some cases may substitute the vinegar with other products like lemon juice or citric acid, is known in Germany as *Tafelmeerrettich.* Also popular in the UK is Tewkesbury mustard, a blend of mustard and grated horseradish originating in medieval times and mentioned by Shakespeare (Falstaff says: "his wit's as thick as Tewkesbury Mustard" in Henry IV Part II). A very similar mustard, called *Krensenf* or *Meerrettichsenf*, is popular in Austria and parts of Eastern Germany.

In the U.S., the term "horseradish sauce" refers to grated horseradish combined with mayonnaise or salad dressing. Prepared horseradish is a common ingredient in Bloody Mary cocktails and in cocktail sauce, and is used as a sauce or sandwich spread.

The distinctive pungent taste of horseradish is from the compound allyl isothiocyanate. Upon crushing the flesh of horseradish, the enzyme myrosinase is released and acts on the glucosinolates sinigrin and gluconasturtiin, which are precursors to the allyl isothiocyanate. The allyl isothiocyanate serves the plant as a natural defense against herbivores. Since allyl isothiocyanate is harmful to the plant itself, it is stored in the harmless form of the glucosinolate, separate from the myrosinase enzyme. When an animal chews the plant, the allyl isothiocyanate is released, repelling the animal. Allyl isothiocyanate is an unstable compound, degrading over the course of days at 37 °C. Because of this instability, horseradish sauces lack the pungency of the freshly crushed roots.

Vegetable

In Central and Eastern Europe horseradish is called *khreyn* (in various spellings like *kren*) in many Slavic languages, in Austria, in parts of Germany (where the other German name *Meerrettich* isn't used), in North-East Italy, and in Yiddish (*ëøééï* translitered as *khreyn*).

There are two varieties of khreyn. "Red" khreyn is mixed with red beet (beetroot) and "white" khreyn contains no beet. It is popular in Ukraine (under the name of *õð³í, khrin*), in Poland (under the name of *chrzan*), in Lithuania (krienai) in the Czech Republic (*køen*), in Russia (*õðåí, khren*), in Hungary (*torma*), in Romania (*hrean*), in Bulgaria (*õðÿí, khryan*), and in Slovakia (under the name of *chren*). Having this on the table is a part of Christian Easter and Jewish Passover tradition in Eastern and Central Europe.

A variety with red beet is called *æwik³a z chrzanem* or simply æwik³a in Poland. In Ashkenazi European Jewish cooking beet horseradish is commonly served with gefilte fish. Red beet with horseradish is also used as a salad served with lamb dishes at Easter called *sfecla cu hrean* in Transylvania and other Romanian regions. Horseradish (often grated and mixed with cream, hard-boiled eggs, or apples) is also a traditional Easter dish in Slovenia and in the adjacent Italian region of Friuli Venezia Giulia; it is used also in the other nearby Italian region of Veneto. In Croatia freshly grated horseradish (Croatian: Hren) is often eaten with boiled ham or beef. In Serbia *ren* is an essential condiment with cooked meat and freshly roasted suckling pig.

Horseradish is also used as a main ingredient for soups. In the Polish region of Silesia, horseradish soup is a common Easter Day dish.

Relation to Wasabi

The Japanese condiment wasabi, although traditionally prepared from the wasabi plant, is now usually made with horseradish due to the scarcity of the wasabi plant. The Japanese botanical name for horseradish is *seiyôwasabi* (»0£0è0¦0ï0µ0Ó0, n‰mq\u„), or "Western wasabi". Both plants are members of the family Brassicaceae.

Nutritional and Biomedical Uses

Compounds found in horseradish have been widely studied for a variety of health benefits. Horseradish contains volatile oils, notably mustard oil, which has antibacterial properties due to the presence of allyl isothiocyanate. Fresh, the plant also contains average 79.31 mg of vitamin C per 100 g of raw horseradish.

The enzyme horseradish peroxidase (HRP), found in the plant, is used extensively in molecular biology and biochemistry.

Artemisia (Genus)

Artemisia (pron.: /ÌQrthÈmiÐziY/) is a large, diverse genus of plants with between 200 and 400 species belonging to the daisy family

Asteraceae. Common names for various species in the genus include mugwort, wormwood, and sagebrush. *Artemisia* comprises hardy herbaceous plants and shrubs, which are known for the powerful chemical constituents in their essential oils. *Artemisia* species grow in temperate climates of both hemispheres, usually in dry or semiarid habitats.

Notable species include *A. vulgaris* (common mugwort), *A. tridentata* (big sagebrush), *A. annua* (sagewort), *A. absinthum* (wormwood), *A. dracunculus* (tarragon), and *A. abrotanum* (southernwood). The leaves of many species are covered with white hairs.

Most species have strong aromas and bitter tastes from terpenoids and sesquiterpene lactones, which exist as an adaptation to discourage herbivory. The small flowers are wind-pollinated. *Artemisia* species are used as food plants by the larvae of a number of Lepidoptera species.

Some botanists split the genus into several genera, but DNA analysis does not support the maintenance of the genera *Crossostephium*, *Filifolium*, *Neopallasia*, *Seriphidium*, and *Sphaeromeria*; three other segregate genera *Stilnolepis*, *Elachanthemum*, and *Kaschgaria*, are maintained by this evidence. Occasionally, some of the species are called sages, causing confusion with the *Salvia* sages in the family Lamiaceae.

Name

The name "artemisia" ultimately derives from the Greek goddess Artemis (Roman Diana), the namesake of Greek Queens Artemisia I and II. A more specific reference may be to Artemisia II of Caria, a botanist and medical researcher who died in 350 BC.

Cultivation and Uses

The aromatic leaves of many species of *Artemisia* are medicinal, and some are used for flavouring. Most species have an extremely bitter taste. *A. dracunculus* (tarragon) is widely used as a culinary herb, particularly important in French cuisine.

Artemisia absinthium (absinth wormwood) was used to repel fleas and moths, and in brewing (wormwood beer, wormwood wine). The aperitif vermouth (derived from the German word *Wermut*, "wormwood") is a wine flavored with aromatic herbs, but originally with wormwood. The highly potent spirits absinthe and Malört also contain wormwood, and the Polish vodka Zoladkowa Gorzka is flavoured with it. Wormwood has been used medicinally as a tonic, stomachic, febrifuge and anthelmintic.

Artemisia arborescens (tree wormwood, or *sheeba* in Arabic) is a very bitter herb indigenous to the Middle East used in tea, usually with mint. In small quantities (in tea), it is believed to have medicinal properties, pacifying various kinds of digestion turmoils. In Israel, *Artemisia* is sometimes referred to as *Shiva*, the Queen of Sheba.

Within Wicca, both wormwood and mugwort are believed to have effects on psychic abilities. Because of the power believed to be inherent in certain herbs of the genus *Artemisia*, many believers cultivate the plants in a "moon garden".

The beliefs surrounding this genus are founded upon the strong association between the herbs of the genus *Artemisia* and the moon goddess Artemis, who is believed to hold these powers.

The bitterness of the plant led to its use by wet-nurses for weaning infants from the breast, as in this speech by Shakespeare from *Romeo and Juliet* Act I, Scene 3:

Nurse: ...

And she [Juliet] was wean'd,–I never shall forget it, –

Of all the days of the year, upon that day:

For I had then laid wormwood to my dug,

...

Shakespeare also refers to the herb as "Dian's bud" (Diana being the Roman incarnation of Artemis) in *Midsummer Night's Dream*, as the antidote to the love potion concocted from the flower "Love in Idleness" (the pansy) that Oberon and Puck use to enchant the lovers:

Oberon: ...

Be as thou wast wont to be:

See as thou wast wont to see:

Dian's bud o'er Cupid's flower

Hath such force and blessed power.

MND 4.1.70-73

...

A few species are grown as ornamental plants, the fine-textured ones used for clipped bordering. All grow best in free-draining sandy soil, unfertilized, and in full sun.

Artemisinin (from *Artemisia annua*) is the active ingredient in the antimalarial combination therapy Coartem produced by Novartis and the World Health Organization.

Artemesia stelleriana is known as 'Dusty Miller', but several other species bear that name, including *Senecio cineraria, Lychnis coronaria,* and *Centaurea cineraria.*

Medicinal

Artemisia oils had inhibitory effects on the growth of bacteria (*Escherichia coli, Staphylococcus aureus*, and *Staphylococcus epidermidis*), yeasts (*Candida albicans, Cryptococcus neoformans*), and dermatophytes (*Trichophyton rubrum, Microsporum canis*, and *Microsporum gypseum*), *Fonsecaea pedrosoi* , and *Aspergillus niger.*

Artemisinin and its derivatives are a group of drugs that possess the most rapid action of all current drugs against malaria. Treatments containing an artemisinin derivative (artemisinin-combination therapies) are now standard treatment worldwide for *P. falciparum* malaria. The starting compound artemisinin is isolated from the plant *Artemisia annua.*

Wormwood was a common ingredient in old abortion recipes. Colette said, "My dream took place under the plant of ill-fame, wormwood ... the terrible age-old inducer of abortions ....".

Artemisia in Popular Culture

Artemisia has been mentioned and used in popular culture for centuries. A few examples are:

- Wormwood (*apsinthos* in the Greek text) is the "name of the star" in the Book of Revelation (8:11) (*kai to onoma tou asteros legetai ho Apsinthos*) that John the Evangelist envisions as cast by the angel and falling into the waters, making them undrinkably bitter. Further references in the Bible show wormwood was a common herb and its awful taste was known. (Deuteronomy 29:18, Book of Proverbs 5:4, Jeremiah 9:15, 25:15, Lamentations 3:15,19 Amos 5:7)
- In the Roald Dahl novel *Matilda* and its 1996 movie adaptation, Matilda's last name is Wormwood.
- In Shakespeare's *Hamlet*, the titular character says "Wormwood, wormwood" to comment on the figurative bitterness of what someone has just said.
- "Wormwood" is a junior devil in *The Screwtape Letters*, a novel by C. S. Lewis on human temptation.
- "Miss Wormwood" is the name of Calvin's teacher in *Calvin and Hobbes*, a former daily comic strip by Bill Watterson. This character is named after the *Screwtape Letters* character.

- In Russian culture, because *Artemisia* species are commonly used in medicine, and their bitter taste is associated with medicinal effects, wormwood is seen as a symbol for a "bitter truth" that must be accepted by a deluded (often self-deluded) person.

  This symbol has acquired a particular poignancy in modern Russian poetry, which often deals with the loss of illusory beliefs in various ideologies.
- Fort Collins, Colorado-based New Belgium brewery produced a spring ale called "Springboard" containing wormwood, *Lycium*, and *Schisandra*.
- In A Series of Unfortunate Events, there is a saying that "tea should be as bitter as wormwood and sharp as a two-edged sword".
- In the Harry Potter series, the Draught of Living Death, an extremely powerful sleeping potion, is made from powdered root of asphodel added to an infusion of wormwood.
- In the video game "Far Cry 2", although the player character is sick with malaria, he is forced to destroy artemisia plants by a local paramilitary faction.

Selected species

Artemisia abrotanum L.–Southernwood, Southern Wormwood, Abrotanum, Lemon Plant

Artemisia absinthium L.–Grand Wormwood

Artemisia adamsii Besser

Artemisia afra Jacq. ex Willd.–African Wormwood, African Sagebrush

Artemisia alaskana Rydb.–Alaska Wormwood

Artemisia alcockii Pamp.

Artemisia aleutica Hultén–Aleutian Wormwood

Artemisia amoena Poljakov

Artemisia annua L.–Annual Wormwood, Sweet Sagewort, Sweet Annie

Artemisia araxina Takht.

Artemisia arborescens–Tree Wormwood

Artemisia arbuscula Nutt.–Little Sagebrush, Low Sagebrush, Black Sage

Artemisia arctica Less.–Boreal Sagebrush

Artemisia arctisibirica Korobkov

Artemisia arenaria DC.

Artemisia arenicola Krasch. ex Poljakov

Artemisia argentata Klokov

Artemisia argentea L'Hér.–Madeira wormwood

Artemisia argyi H.Lév. & Vaniot

Artemisia argyrophylla Ledeb.

Artemisia armeniaca Lam.

Artemisia aschurbajewii C.G.Aro

Artemisia australis Less.–»Âhinahina, O»ahu Wormwood

Artemisia austriaca Jacq.

Artemisia avarica Minat.

Artemisia badhysi Krasch. & Lincz. ex Poljakov

Artemisia balchanorum Krasch.

Artemisia baldshuanica Krasch. & Zaprjag.

Artemisia bargusinensis Spreng.

Artemisia bejdemaniae Leonova

Artemisia biennis Willd.–Biennial Sagewort, Biennial Wormwood

Artemisia bigelovii A.Gray–Bigelow Sage, Bigelow Sagebrush

Artemisia borealis Pall.

Artemisia borotalensis Poljakov

Artemisia bottnica Lundstr. ex Kindb.

Artemisia caespitosa Ledeb.

Artemisia californica Less.–Coastal Sagebrush, California Sagebrush

Artemisia camelorum Krasch.

Artemisia campestris L.–Field Wormwood

Artemisia camphorata Vill.

Artemisia cana Pursh–Silver Sagebrush

Artemisia canadensis Michx.–Canada Wormwood

Artemisia capillaris Thunb.–Capillary Wormwood

Artemisia carruthii Wood ex Carruth.–Carruth Sagewort, Carruth's Sagebrush

Artemisia caruifolia Buch.-Ham. ex Roxb.

Artemisia caucasica Willd.

Artemisia chamaemelifolia Vill.

Artemisia cina O.Berg & C.F.Schmidt–Santonica, Levant Wormseed

Artemisia ciniformis Krasch. & Popov ex Poljakov

Artemisia commutata Besser

Artemisia compacta Fisch. ex DC.

Artemisia cuspidata Krasch.

Artemisia czukavinae Filatova

Artemisia daghestanica Krasch. & Poretzky

Artemisia demissa Krasch.

Artemisia depauperata Krasch.

Artemisia deserti Krasch.

Artemisia desertorum Spreng.

Artemisia diffusa Krasch. ex Poljakov

Artemisia dimoana Popov

Artemisia dolosa Krasch.

Artemisia douglasiana Bess.–Douglas' Mugwort, Douglas' Sagewort

Artemisia dracunculus L.–Tarragon

Artemisia dubia Wall.

Artemisia dubjanskyana Krasch. ex Poljakov

Artemisia dumosa Poljakov

Artemisia elongata Filatova & Ladygina

Artemisia eremophila Krasch. & Butkov ex Poljakov

Artemisia eriantha Ten.

Artemisia feddei H.Lév. & Vaniot

Artemisia fedtschenkoana Krasch.

Artemisia ferganensis Krasch. ex Poljakov

Artemisia filifolia Torr.–Sand Sagebrush, Sand Sagebush, Silvery Wormwood

Artemisia flava Jurtzev

Artemisia franserioides Greene–Ragweed Sagebrush

Artemisia freyniana (Pamp.) Krasch.

Artemisia frigida Willd.–Fringed Sagebrush, Fringed Sagewort, Prairie Sagewort

Artemisia fulvella Filatova & Ladygina

Artemisia furcata Bieb.–Forked Wormwood

Artemisia galinae Ikonn.

Artemisia genipi Weber ex Stechm.

Artemisia glabella Kar. & Kir.

Artemisia glacialis L.–Glacier Wormwood, Alpine Mugwort

Artemisia glanduligera Krasch. ex Poljakov

Artemisia glauca Pall. ex Willd.

Artemisia glaucina Krasch. ex Poljakov

Artemisia globosa Krasch.

Artemisia globularia Cham. ex Bess.–Purple Wormwood

Artemisia glomerata Ledeb.–Cudweed Sagewort, Pacific Alpine Wormwood

Artemisia gmelinii Webb ex Stechmann–Gmelin's Wormwood

Artemisia gnaphalodes Nutt.

Artemisia gorjaevii Poljakov

Artemisia gracilescens Krasch. & Iljin

Artemisia granatensis Boiss. ex DC.

Artemisia gurganica (Krasch.) Filatova

Artemisia gypsacea Krasch., Popov & Lincz. ex Poljakov

Artemisia halodendron Turcz. ex Besser

Artemisia halophila Krasch.

Artemisia heptapotamica Poljakov

Artemisia herba-alba Asso–White Wormwood

Artemisia hippolyti Butkov

Artemisia hololeuca M.Bieb. ex Besser

Artemisia hulteniana Vorosch.

Artemisia incana (L.) Druce

Artemisia indica Willd.–Yomogi

Artemisia insulana Krasch.

Artemisia insularis Kitam.

Artemisia integrifolia L.

Artemisia issykkulensis Poljakov

Artemisia jacutica Drobow

Artemisia japonica Thunb.–Otoko Yomogi

Artemisia juncea Kar. & Kir.

Artemisia karatavica Krasch. & Abolin ex Poljakov

Artemisia karavajevii Leonova

Artemisia kaschgarica Krasch.

Artemisia kauaiensis (Skottsberg) Skottsberg–»Âhinahina, Kaua»i Wormwood

Artemisia keiskeana Miq.

Artemisia kelleri Krasch.

Artemisia kemrudica Krasch.

Artemisia knorringiana Krasch.

Artemisia kochiiformis Krasch. & Lincz. ex Poljakov

Artemisia koidzumii Nakai

Artemisia kopetdaghensis Krasch. ex Poljakov

Artemisia korovinii Poljakov

Artemisia korshinskyi Krasch. ex Poljakov

Artemisia krushiana Bess.–Krush's Wormwood

Artemisia kulbadica Boiss. & Buhse

Artemisia kuschakewiczii C.G.A.Winkl.

Artemisia laciniata Willd.–Siberian Wormwood

Artemisia laciniatiformis Kom.

Artemisia lactiflora Kom.

Artemisia lagocephala (Besser) DC.

Artemisia lagopus Fisch. ex Besser

Artemisia lanata Willd.

Artemisia latifolia Ledeb.

Artemisia ledebouriana Besser

Artemisia lehmanniana Bunge

Artemisia leontopodioides Fisch. ex Besser

Artemisia lessingiana Besser

Artemisia leucodes Schrenk

Artemisia leucophylla (Turcz. ex Besser) Pamp.

Artemisia leucotricha Krasch. ex Ladygina

Artemisia lindleyana Bess.–Columbia River Wormwood

Artemisia lipskyi Poljakov

Artemisia littoricola Kitam.

Artemisia longifolia Nutt.–Longleaf Sagebrush, Longleaf Wormwood

Artemisia ludoviciana Nutt.–Gray Sagewort, Prairie Sage, White Sagebrush

Artemisia macilenta (Maxim.) Krasch.

Artemisia macrantha Ledeb.

Artemisia macrobotrys Ledeb.–Yukon Wormwood

Artemisia macrocephala Jacq. ex Besser

Artemisia macrorhiza Turcz.

Artemisia maracandica Bunge

Artemisia maritima L.–Sea Wormwood

Artemisia marschalliana Spreng.

Artemisia martjanovii Krasch. ex Poljakov

Artemisia mauiensis (A.Gray) Skottsberg–»Âhinahina, Maui Wormwood

Artemisia maximovicziana Krasch. ex Poljakov

Artemisia medioxima Krasch. ex Poljakov

Artemisia messerschmidtiana Besser

Artemisia michauxiana Bess.–Michaux Sagebrush, Michaux's Wormwood

Artemisia minor Jacq. ex Besser

Artemisia mogoltavica Poljakov

Artemisia mongolica (Besser) Fisch. ex Nakai

Artemisia mongolorum Krasch.

Artemisia montana (Nakai) Pamp.

Artemisia mucronulata Poljakov

Artemisia multisecta Leonova

Artemisia mutellina Vill.

Artemisia nachitschevanica Rzazade

Artemisia nakaii Pamp.

Artemisia namanganica Poljakov

Artemisia nana Gaudin

Artemisia negrei Ouyahya

Artemisia nesiotica Raven–Island Sagebrush

Artemisia nigricans Filatova & Ladygina

Artemisia niitakayamensis Hayata

Artemisia nilagirica (C.B.Clarke) Pamp.

Artemisia nitida Bertol.

Artemisia nortonii Pamp.

Artemisia norvegica Fr.–Norwegian Mugwort

Artemisia nova A.Nels.–Black Sagebrush

Artemisia nuristanica Kitam.

Artemisia obscura Pamp.

Artemisia obtusa Rydb.

Artemisia obtusiloba Ledeb.

Artemisia occidentalisichuanensis Y.R.Ling & S.Y.Zhao

Artemisia occidentalisinensis Y.R.Ling

Artemisia oelandica (Besser) Krasch.

Artemisia olchonensis Leonova

Artemisia oliveriana J.Gay ex Besser

Artemisia ordosica Krasch.

Artemisia orientalixizangensis Y.R.Ling & Humphries

Artemisia orientaliyunnanensis Y.R.Ling

Artemisia orthobotrys Kitag.

Artemisia packardiae J.Grimes & Ertter–Packard's Wormwood, Succor Creek Sagebrush

Artemisia pallasiana Fisch. ex Besser

Artemisia palmeri A.Gray–San Diego Sagewort

Artemisia palustris L.

Artemisia pannosa Krasch.

Artemisia papposa S.F.Blake & Cronq.–Owyhee Sage, Owyhee Sagebrush

Artemisia parryi A.Gray–Parry's Wormwood

Artemisia pattersonii A.Gray–Patterson's Wormwood

Artemisia pectinata Pall.

Artemisia pedatifida Nutt.–Birdfoot Sagebrush

Artemisia pedemontana Balb.

Artemisia persica Boiss.

Artemisia pewzowii C.G.A.Winkl.

Artemisia phaeolepis Krasch.

Artemisia polysticha Poljakov

Artemisia pontica L.–Roman Wormwood

Artemisia porrecta Krasch. ex Poljakov

Artemisia porteri Cronq.–Porter's Wormwood

Artemisia prasina Krasch. ex Poljak.

Artemisia princeps Pamp.–Japanese Mugwort, Yomogi

Artemisia proceriformis Krasch.

Artemisia prolixa Krasch. ex Poljak.

Artemisia punctigera Krasch. ex Poljakov

Artemisia purshiana Besser

Artemisia pycnocephala (Less.) DC.–Beach Wormwood

Artemisia pycnorhiza Ledeb.

Artemisia pygmaea A.Gray–Pygmy Sagebrush

Artemisia quinqueloba Trautv.

Artemisia remotiloba Krasch. ex Poljakov

Artemisia rhodantha Rupr.

Artemisia rigida (Nutt.) A.Gray–Scabland Sagebrush

Artemisia rothrockii A.Gray–Timberline Sagebrush

Artemisia roxburghiana Wall. ex Besser

Artemisia rubripes Nakai

Artemisia rupestris L.–Rock Wormwood

Artemisia rutifolia Stephan ex Spreng.

Artemisia sacrorum Ledeb. ex Hook.f.

Artemisia saissanica (Krasch.) Filatova

Artemisia saitoana Kitam.

Artemisia salsoloides Willd.

Artemisia samoiedorum Pamp.

Artemisia santolina Schrenk

Artemisia santolinifolia Turcz. ex Besser

Artemisia santonica L.

Artemisia saposhnikovii Krasch. ex Poljak.

Artemisia schischkinii Krasch.

Artemisia schmidtiana

Artemisia schrenkiana Ledeb.

Artemisia scoparia Waldst. & Kit.–Redstem Wormwood

Artemisia scopiformis Ledeb.

Artemisia scopulorum A.Gray–Alpine Sagebrush, Dwarf Sagebrush

Artemisia scotina Nevski

Artemisia senjavinensis Bess.–Arctic Wormwood

Artemisia semiarida (Krasch. & Lavrenko) Filatova

Artemisia senjavinensis Besser

Artemisia sericea Weber ex Stechm.

Artemisia serotina Bunge

Artemisia serrata Nutt.–Sawtooth Wormwood

Artemisia sieversiana Willd.

Artemisia skorniakowii C.G.A.Winkl.

Artemisia sogdiana Bunge

Artemisia songarica Schrenk

Artemisia spicigera K.Koch

Artemisia spinescens D.C.Eaton–Budsage
[= *Picrothamnus desertorum*]

Artemisia splendens Willd.

Artemisia stelleriana Bess.–Hoary Mugwort or Dusty Miller (one of several plants with this name)

Artemisia stenocephala Krasch. ex Poljak.

Artemisia stenophylla Kitam.

Artemisia stolonifera (Maxim.) Kom.

Artemisia subarctica Krasch.

Artemisia subchrysolepis Filatova

Artemisia sublessingiana Krasch. ex Poljakov

Artemisia subsalsa Filatova

Artemisia subviscosa Turcz. ex Besser

Artemisia succulenta Ledeb.

Artemisia suksdorfii Piper–Coastal Wormwood

Artemisia sylvatica Maxim.

Artemisia szowitziana (Besser) Grossh.

Artemisia tanacetifolia L.

Artemisia taurica Willd.

Artemisia tenuisecta Nevski

Artemisia terrae-albae Krasch.

Artemisia tianschanica Krasch. ex Poljak.

Artemisia tilesii Ledeb.–Tilesius' Wormwood

Artemisia tomentella Trautv.

Artemisia tournefortiana Rchb.

Artemisia transbaicalensis Leonova

Artemisia transiliensis Poljakov

Artemisia trautvetteriana Besser

Artemisia tridentata Nutt.–Big Sagebrush, Blue Sage, Black Sage, Basin Sagebrush

Artemisia triniana Besser

Artemisia tripartita Rydb.–Threetip Sagebrush

Artemisia turanica Krasch.

Artemisia turcomanica Gand.

Artemisia umbelliformis Lam.–Alps Wormwood

Artemisia unalaskensis Rydb.

Artemisia underwoodii Rydb.

Artemisia uralensis Spreng. ex Besser

Artemisia uraorum Hultén

Artemisia uzbekistanica Poljakov

Artemisia vachanica Krasch. ex Poljak.

Artemisia valida Krasch. ex Poljak.

Artemisia verlotiorum Lamotte–Chinese Wormwood

Artemisia viridis Willd.

Artemisia vulgaris L.–Mugwort

Artemisia wallichiana Besser

Artemisia waltonii J.R.Drumm. ex Pamp.

Artemisia wudanica Liou & W.Wang

Artemisia wulingshanensis Bar. & Skv. ex Liou

Artemisia wurzellii C.M.James & Stace

Artemisia xerophila Magnier

Artemisia xerophytica Krasch.

Artemisia xylorhiza Krasch. ex Filatova

Artemisia yadongensis Ling & Y.R.Ling

Artemisia yongii Y.R.Ling

Artemisia younghusbandii J.R.Drumm. ex Pamp.

Artemisia zayuensis Y.R.Ling

Artemisia zhaodongensis G.Y.Chang & M.Y.Liou

Artemisia zhongdianensis Y.R.Ling

Artemisia zollingeriana Sch.Bip.

Formerly placed here

- *Centipeda minima* (L.) A.Braun & Asch. (as *A. minima* L.)
- *Eupatorium capillifolium* (Lam.) Small (as *A. capillifolia* Lam.)
- *Filifolium sibiricum* (L.) Kitam. (as *A. sibirica* (L.) Maxim.)
- *Grangea maderaspatana* (L.) Poir. (as *A. maderaspatana* L.)
- *Matricaria discoidea* DC. (as *A. matricarioides* auct.)

Classification

Classification of *Artemisia* is difficult. Divisions of *Artemisia* prior to 2000 into subgenera or sections have not been backed up by molecular data, but much of the molecular data, as of 2006, are not especially strong. The following identified groups do not include all the species in the genus.

Section Tridentatae

Section *Tridentatae* consists of nine to eleven species of shrubs, which are very prominent parts of the flora in western North America.

In some classifications, they are part of the genus or subgenus *Seriphidium*, although they do not seem to be closely related to the Asian *Seriphidium* species.

To be monophyletic, section *Tridentatae* should exclude *Artemisia bigelovii* and *Artemisia palmeri.*

- *Artemisia tridentata*
- *Artemisia cana*
- *Artemisia nova*
- *Artemisia rigida*
- *Artemisia arbuscula*
- *Artemisia longiloba*
- *Artemisia tripartita*
- *Artemisia pygmaea*
- *Artemisia rothrockii*

Section *Tridentatae* includes above species with exception of *Artemisia longiloba*, which is treated as a subspecies of *Artemisia arbuscula*. Section *Nebulae* includes *Artemisia californica, Artemisia nesiotica*, and *Artemisia filifolia.*

Old World Seriphidium

The Old World species which different classifications put into the genus or subgenus *Seriphidium* consist of about 125 species native to Europe and temperate Asia, with the largest number of species in Central Asia. Some classifications, such as that of the Flora of North America, exclude any New World plants from *Seriphidium*. They are herbaceous plants or small shrubs.

Subgenus Dracunculus

One group which is well-supported by molecular data is subgenus *Dracunculus*. It consists of 80 species found in both North America and Eurasia, of which the best-known is perhaps *Artemisia dracunculus*, the spice tarragon.

Asclepias Syriaca

Asclepias syriaca, commonly called Common Milkweed, Butterfly flower, Silkweed, Silky Swallow-wort, Virginia Silkweed, is a herbaceous plant species. It is in the genus *Asclepias*, making it a type of milkweed. This species is native to most of North America east of the Rocky Mountains, excluding the drier parts of the Prairies. It grows in sandy soils and appreciates lots of sunlight. It was one of the earliest North American species described in Cornut's 1635 *Canadensium plantarum historia*. The specific name was reused by Linnaeus due to Cornut's confusion with a species from Asia Minor.

Common milkweed is a herbaceous perennial plant growing to 1–2 m tall from a rhizome. The stem and all parts of the plants produce a white latex when broken. The leaves are opposite, simple broad ovate-lanceolate, 7–25 cm long and 3–12 cm broad, usually with an undulate margin and a red-coloured main vein. They have a very short petiole and a velvety underside.

The flowers are grouped in several spherical umbels with numerous flowers in each umbel. The individual flowers are small, 1–2 cm diametre, perfumed, with five cornate hoods. The seeds are attached to long, white flossy hairs and encased in large follicles.

Uses

The plant's latex contains large quantities of glycosides, making the leaves and seed pods toxic for sheep and other large mammals, and potentially humans (though large quantities of the foul-tasting parts would need to be eaten). The young shoots, young leaves, flower buds and immature fruits are all edible raw.

Concerns about milkweed bitterness and toxicity can be traced back to Euell Gibbons, author of *Stalking the Wild Asparagus* (1962). It is theorized that Gibbons inadvertently prepared dogbane (*Apocynum cannabinum*), a poisonous look-alike instead. He devised a method to remove the bitterness and toxicity by plunging the young shoots into boiling water (not cold) and cooking for one minute, repeating the procedure at least three times to make the plant safe to eat. Gibbons' method was copied from book to book, dominating edible plants literature for forty years. Most modern foragers consider the bitterness/toxicity issue a myth. The plants have no bitterness when tasted raw, and can be cooked like asparagus, with no special processing.

Failed attempts have been made to exploit rubber (from the latex) and fibre (from the seed's floss) production from the plant industrially. The fluffy seed hairs have been used as the traditional background for mounted butterflies. The compressed floss has a beautiful silk-like sheen.

The plant has also been explored for commercial use of its bast (inner bark) fibre which is both strong and soft. U. S. Department of Agriculture studies in the 1890s and 1940s found that Milkweed has more potential for commercial processing than any other indigenous bast fibre plant, with estimated yields as high as hemp and quality as good as flax.

Both the bast fibre and the floss were used historically by Native Americans for cordage and textiles. Milkweed oil from the seeds can be easily converted into cinnamic acid, which is a very potent sunscreen when used at a 1-5% concentration.

The flowers often constitute small traps for insects who cannot take off again. Several insects live off the plant, including the Monarch Butterfly (*Danaus plexippus*), the Milkweed Beetle (*Tetraopes tetraophtalmus*), Large Milkweed Bug (*Oncopeltus fasciatus*), Small Milkweed Bug (*Lygaeus kalmii*), the Milkweed Leaf Beetle (*Labidomera clivicollis*) and the weevil *Rhyssomatus lineaticollis*. The flower nectar has a high glucose content and was used by natives as a sweetener.

Deforestation due to European settlement may have expanded the range and density of milkweed. The plant can become invasive and often acts as a weed. It is naturalized in several areas outside of its native range, including Oregon and parts of Europe.

Astragalus

Astragalus (*As-trá-ga-lus*) is a large genus of about 3,000 species of herbs and small shrubs, belonging to the legume family Fabaceae, subfamily Faboideae. The genus is native to temperate regions of the Northern Hemisphere. Common names include milk-vetch (most species), locoweed (in western US, some species; although most locoweeds are not genus *Astragalus* but in related genera) and goat's-thorn (*A. gummifer*, *A. tragacanthus*). Some pale-flowered vetches are similar in appearance, but vetches are more vine-like.

Species

Astragalus species are used as food plants by the larvae of some Lepidoptera species including the following case-bearers of the genus *Coleophora*: *C. astragalella* (feeds exclusively on *A. glycyphyllos*), *C. cartilaginella* (feeds exclusively on *Astragalus*), *C. colutella*, *C. euryaula* (feeds exclusively on *Astragalus*), *C. gallipennella* (feeds exclusively on *A. glycyphyllos*), *C. hippodromica* (feeds exclusively on *A. gombo*), *C. onobrychiella* (feeds exclusively on *Astragalus*), *C. polonicella* (feeds exclusively on *A. arenarius*) and *C. vicinella*.

Traditional Uses

The natural gum tragacanth is made from several species of *Astragalus* occurring in the Middle East, including *A. adscendens*, *A. gummifer*, *A. brachycalyx*, and *A. tragacanthus*. Also *Astragalus propinquus* (also known as *Astragalus membranaceus*) has a history of use as a herbal medicine and is used in traditional Chinese medicine.

Research

The biotechnology firms Geron Corporation and TA Therapeutics of Hong Kong have been working on deriving a telomerase activator from it. The chemical constituent cycloastragenol (also called TAT2) is being studied to help combat HIV, as well as infections associated with chronic diseases or aging. However, the National Institutes of Health states: "The evidence for using astragalus for any health condition is limited. High-quality clinical trials (studies in people) are generally lacking. There is some preliminary evidence to suggest that astragalus, either alone or in combination with other herbs, may have potential benefits for the immune system, heart, and liver, and as an adjunctive therapy for cancer".

In a study published in the 15 November 2008 *Journal of Immunology*, researchers shed light on the antiviral and antiaging benefits of *Astragalus*. In their paper, a team of researchers from the UCLA AIDS Institute described how their work with cycloastragenol reduces the aging process of immune cells, and enhances how these cells respond to viral infections. The compound works by boosting production of telomerase, an enzyme that allows for the replacement of short bits of DNA, known as telomeres, that play a key role in cell replication, cancer and human aging.

Supplement Use

Extracts of *Astragalus membranaceus* (syn. *Astragalus propinquus*) are marketed as life prolonging extracts in humans. A proprietary extract of the dried root of *Astragalus membranaceus*, called TA-65, 'was associated with a significant age-reversal effect in the immune system, in that it led to declines in the percentage of senescent cytotoxic T cells and natural killer cells after six to twelve months of use.'

In October 2010, Intertek/AAC Labs, an ISO 17025 internationally recognized lab, found the largest component of TA-65 to be Cycloastragenol. Telomerase activation was feared to pose an increased risk of cancer because telomere shortening is a mechanism that limits cell proliferation.

However, short telomeres result in chromosome instability, hence there is also a potential mechanism for telomere lengthening to protect against cancer (as distinct from mutation-induced activation).

Side Effects

Astragalus may interact with medications that suppress the immune system, such as cyclophosphamide. It may also affect blood sugar levels and blood pressure. Some *Astragalus* species can be toxic. For example, several species that grow in the United States contain the neurotoxin swainsonine.

Toxicology: The family of *Astragalus* has been reviewed by Rios and Waterman, indicating that the toxicities vary from one species to another.

Ornamental Use: Several species, including *A. alpinus* (bluish-purple flowers), *A. hypoglottis* (purple flowers) and *A. lotoides*, are grown as ornamental plants in gardens.

Selected Species: *This list is incomplete; you can help by expanding it.*

- *Astragalus acutirostris* – sharpkeel milkvetch
- *Astragalus agnicidus* – Humboldt County milkvetch
- *Astragalus agrestis* – field milkvetch, purple milkvetch, cock's-head
- *Astragalus aitosensis*
- *Astragalus albens* – Cushenbury milkvetch
- *Astragalus albertii*
- *Astragalus alopecuroides*
- *Astragalus alpinus* – alpine milkvetch, mountain locoweed
- *Astragalus amphioxys* – crescent milkvetch
- *Astragalus ampullarioides* – Shivwits milkvetch
- *Astragalus andersonii* – Anderson's milkvetch
- *Astragalus angustifolius*
- *Astragalus anisus*
- *Astragalus annularis*
- *Astragalus anserinus* – Goose Creek milkvetch
- *Astragalus anxius* – troubled milkvetch
- *Astragalus applegatei* – Applegate's milkvetch
- *Astragalus arenarius*

- *Astragalus aristatus*
- *Astracantha arnacantha*
- *Astragalus asymmetricus* – San Joaquin milkvetch
- *Astragalus atratus* – Owyhee mourning milkvetch
- *Astragalus atropilosulus*
- *Astragalus austiniae* – Austin's milkvetch
- *Astragalus australis* - Indian milkvetch
- *Astragalus azizi*
- *Astragalus baionensis*
- *Astragalus balearicus* – Balearic milkvetch
- *Astragalus barrii* – Barr's milkvetch
- *Astragalus bernardinus* – San Bernardino milkvetch
- *Astragalus bibullatus* – limestone-glade milkvetch
- *Astragalus bicristatus* – crested milkvetch, two-crested milkvetch
- *Astragalus bidentatus*
- *Astragalus bisulcatus* – two-groove milkvetch
- *Astragalus boeticus*
- *Astragalus bolanderi* – Bolander's milkvetch
- *Astragalus brachycalyx*
- *Astragalus brauntonii* – Braunton's milkvetch
- *Astragalus breweri* – Brewer's milkvetch
- *Astragalus californicus* – California milkvetch
- *Astragalus canadensis* – Canadian milkvetch
- *Astragalus casei* – Case's milkvetch
- *Astragalus centralpinus*
- *Astragalus christianus*
- *Astragalus cicer* – wild lentil, chickpea milkvetch
- *Astragalus cimae* – Cima milkvetch
- *Astragalus clarianus* syn. *Astragalus claranus* – Clara Hunt's milkvetch, Napa milkvetch
- *Astragalus clevelandii* – Cleveland's milkvetch
- *Astragalus coccineus* – scarlet milkvetch, scarlet locoweed
- *Astragalus congdonii* – Congdon's milkvetch

- *Astragalus crassicarpus* (*caryocarpus*) – ground plum, buffalo plum
- *Astragalus cremnophylax* – Sentry milkvetch
- *Astragalus crotalariae* – Salton milkvetch
- *Astragalus curtipes* – Morro milkvetch
- *Astragalus danicus*
- *Astragalus deanei* – Deane's milkvetch, Dean's milkvetch
- *Astragalus debequaeus* – Debeque milkvetch
- *Astragalus depressus*
- *Astragalus desereticus* – Deseret milkvetch
- *Astragalus didymocarpus* – two-seeded milkvetch, dwarf white milkvetch, white dwarf locoweed
- *Astragalus douglasii* – Douglas's milkvetch
- *Astragalus echinus*
- *Astragalus eriocarpus*
- *Astragalus ertterae* – Walker Pass milkvetch
- *Astragalus exscapus*
- *Astragalus filipes* – basalt milkvetch
- *Astragalus frigidus* – American milk-vetch, arctic milkvetch
- *Astragalus funereus* – Funeral Mountain milkvetch
- *Astragalus galegiformis*
- *Astragalus gambelianus* – Gambel's dwarf milkvetch, dwarf locoweed
- *Astragalus gibbsii* – Gibbs' milkvetch
- *Astragalus gilmanii* – Gilman's milkvetch
- *Astragalus gilviflorus* – plains milkvetch, Plains orophaca
- *Astragalus glycyphyllos* – wild liquorice, licorice milkvetch
- *Astraglus gossypinus*
- *Astragalus gummifer*
- *Astragalus holmgreniorum* – Holmgren locoweed
- *Astragalus humillimus* – Mancos milkvetch
- *Astragalus hypoglottis* – field milkvetch, purple milkvetch, cock's-head
- *Astragalus hypoxylus* - Huachuca Mountain milkvetch
- *Astragalus illyricus* – Illirian milkvetch

- *Astragalus inversus* – Susanville milkvetch
- *Astragalus inyoensis* – Inyo milkvetch
- *Astragalus iselyi* - Isely's milkvetch
- *Astragalus jaegerianus* – Lane Mountain milkvetch
- *Astragalus johannis-howellii* – Long Valley milkvetch
- *Astragalus kentrophyta* – spiny milkvetch
- *Astragalus layneae* – widow's milkvetch
- *Astragalus lemmonii* – Lemmon's milkvetch
- *Astragalus lentiformis* – lens pod milkvetch
- *Astragalus lentiginosus* – freckled milkvetch, mottled locoweed, speckled locoweed, spotted locoweed
- *Astragalus leontinus*
- *Astragalus leptaleus*
- *Astragalus leucolobus* – Bear Valley woollypod, Bear Valley milkvetch
- *Astragalus linifolius* - Grand Junction milkvetch
- *Astragalus loanus* - Glenwood milkvetch
- *Astragalus lotoides*
- *Astragalus lusitanicus*
- *Astragalus malacus* – shaggy milkvetch
- *Astragalus massiliensis*
- *Astragalus microcymbus* - Skiff milkvetch
- *Astragalus miguelensis* – San Miguel milkvetch
- *Astragalus missouriensis* – Missouri milkvetch
- *Astragalus mohavensis* – Mojave milkvetch
- *Astragalus mollissimus* - woolly locoweed, stemmed locoweed
- *Astragalus molybdenus*
- *Astragalus monoensis* – Mono milkvetch
- *Astragalus monspessulanus*
- *Astragalus montii*
- *Astragalus mulfordiae* - Mulford's milkvetch
- *Astragalus naturitensis* – Naturita milkvetch
- *Astragalus nevinii* – San Clemente Island milkvetch
- *Astragalus newberryi* – Newberry's milkvetch

- *Astragalus nitidiflorus* – Tallante's milkvetch
- *Astragalus norvegicus*
- *Astragalus nutans* – Providence Mountains milkvetch
- *Astragalus nuttallianus* – small-flowered milkvetch
- *Astragalus nuttallii* – Nuttall's milkvetch
- *Astragalus obscurus* – arcane milkvetch
- *Astragalus onobrychis*
- *Astragalus oocarpus* – Descanso milkvetch, San Diego milkvetch
- *Astragalus oophorus* – egg milkvetch
- *Astragalus osterhoutii* – Osterhout milkvetch
- *Astragalus oxyphysus* – Mt. Diablo milkvetch, Diablo locoweed
- *Astragalus pachypus* – thickpod milkvetch
- *Astragalus panamintensis* – panamint milkvetch
- *Astragalus pauperculus* – depauperate milkvetch
- *Astragalus penduliflorus*
- *Astragalus phoenix* – Ash Meadows milkvetch
- *Astragalus platytropis* – broadkeel milkvetch
- *Astragalus polycladus*
- *Astragalus pomonensis* – Pomona milkvetch, Pomona locoweed
- *Astragalus propinquus* (syn. *A. membranaceus*) – *huang qi*
- *Astragalus proximus*
- *Astragalus pseudiodanthus* – Tonopah milkvetch
- *Astragalus pulsiferae* – Ames' milkvetch
- *Astragalus purshii* – Pursh's milkvetch, Woollypod milkvetch, woollypod locoweed
- *Astragalus pycnostachyus* – Marsh milkvetch
- *Astragalus rattanii* – Rattan's milkvetch
- *Astragalus ripleyi*
- *Astragalus robbinsii* – Robbins' milkvetch
- *Astragalus sabulonum* – gravel milkvetch
- *Astragalus scaphoides* – bitterroot milkvetch
- *Astragalus sclerocarpus* – sicklepod milkvetch
- *Astragalus schmolliae* - Schmoll milkvetch
- *Astragalus sempervirens*

- *Astragalus shevockii* – Shevock's milkvetch, Little Kern milkvetch
- *Astragalus shinanensis*
- *Astragalus shiroumaensis*
- *Astragalus sinicus* – Chinese milkvetch, *zi yun ying* (+}ò–ñ,)
- *Astragalus sinuatus* - Whited's milkvetch
- *Astragalus sirinicus*
- *Astragalus spatulatus* – tufted milk-vetch, Draba milkvetch
- *Astragalus speirocarpus* – coilpod locoweed
- *Astragalus spinosus* – spiny milkvetch
- *Astragalus subvestitus* – Kern County milkvetch
- *Astragalus tener* – alkali milkvetch
- *Astragalus tennesseensis* – Tennessee milk vetch
- *Astragalus tidestromii* – Tidestrøm's milkvetch
- *Astragalus traskiae* – Trask's milkvetch
- *Astragalus tricarinatus* – triple-ribbed milkvetch
- *Astragalus trichopodus* – Santa Barbara milkvetch, coast locoweed, Southern California locoweed
- *Astragalus tridactylus* – foothill milkvetch
- *Astragalus tragacanthus*
- *Astragalus tyghensis*
- *Astragalus umbraticus* – Bald Mountain milkvetch
- *Astragalus vesicarius*
- *Astragalus vogelii*
- *Astragalus webberi* – Webber's milkvetch
- *Astragalus wetherillii*
- *Astragalus whitneyi* – balloon-pod milkvetch
- *Astragalus zionis* – Zion milkvetch

Atropa Belladonna

Atropa belladonna or *Atropa bella-donna*, commonly known as Belladonna or Deadly Nightshade, is a perennial herbaceous plant in the family Solanaceae, native to Europe, North Africa, and Western Asia. The foliage and berries are extremely toxic, containing tropane alkaloids. These toxins include scopolamine and hyoscyamine which cause a bizarre delirium and hallucinations, and are also used as pharmaceutical anticholinergics. The drug atropine is derived from the plant. It has a long history of use as a medicine, cosmetic, and poison. Before the Middle Ages, it was used as an anesthetic for surgery; the ancient Romans used it as a poison (the wife of Emperor Augustus and the wife of Claudius both were rumored to have used it to murder contemporaries); and predating this, it was used to make poison-tipped arrows. The genus name "atropa" comes from Atropos, one of the three Fates in Greek mythology, and the name "bella donna" is derived from Italian and means "beautiful woman" because the herb was used in eye-drops by women to dilate the pupils of the eyes to make them appear seductive.

Description

Atropa belladonna is a branching herbaceous perennial, often growing as a subshrub, from a fleshy rootstock. Plants grow to 1.5 metres (4.9 ft) tall with 18 centimetres (7.1 in) long ovate leaves. The bell-shaped flowers are tyrian purple with green tinges and faintly scented. The fruits are berries, which are green ripening to a shiny black, and approximately 1 centimetre (0.39 in) in diametre. The berries are sweet and are consumed by animals that disperse the seeds in their droppings, even though the seeds contain toxic alkaloids. There is a pale yellow flowering form called *Atropa belladonna* var. *lutea* with pale yellow fruit.

Atropa belladona is rarely used in gardens, but when grown, it is usually for its large upright habit and showy berries. It is naturalized in parts of North America, where it is often found in shady, moist locations with limestone-rich soils. It is considered a weed species in parts of the world, where it colonizes areas with disturbed soils. Germination of the small seeds is often difficult, due to hard seed coats that cause seed dormancy. Germination takes several weeks under alternating temperature conditions, but can be sped up with the use of gibberellic acid. The seedlings need sterile soil to prevent damping off and resent root disturbance during transplanting. This plant is a sign of water near by.

Naming and Taxonomy

The name *Atropa belladonna* was published by Linnaeus in *Species Plantarum* in 1753. It is in the nightshade family (*Solanaceae*), which it shares with potatoes, tomatoes, eggplants, jimsonweed, tobacco, wolfberry, and chili peppers. The common names for this species include belladonna, deadly nightshade, divale, dwale, banewort, devil's berries, naughty man's cherries, death cherries, beautiful death, devil's herb, great morel, and dwayberry. The name *Atropa* is thought to be derived from that of the Greek goddess Atropos, one of the three Greek fates or destinies who would determine the course of a man's life by the weaving of threads that symbolized his birth, the events in his life and finally his death; with Atropos cutting these threads to mark the last of these.

The name "belladonna" comes from the Italian language, meaning "beautiful lady"; originating either from its usage as cosmetic for the face, or, more probably, from its usage to increase the pupil size in women.

Toxicity

Belladonna is one of the most toxic plants found in the Eastern Hemisphere. All parts of the plant contain tropane alkaloids. The berries pose the greatest danger to children because they look attractive and have a somewhat sweet taste. The consumption of two to five berries by a human adult is probably lethal. The root of the plant is generally the most toxic part, though this can vary from one specimen to another. Ingestion of a single leaf of the plant can be fatal to an adult.

The active agents in belladonna, atropine, hyoscine (scopolamine), and hyoscyamine, have anticholinergic properties. The symptoms of belladonna poisoning include dilated pupils, sensitivity to light, blurred vision, tachycardia, loss of balance, staggering, headache, rash, flushing, severely dry mouth and throat, slurred speech, urinary retention, constipation, confusion, hallucinations, delirium, and convulsions.

In 2009, *A. belladonna* berries were mistaken for blueberries by an adult woman; the six berries she ate were documented to result in severe anticholinergic syndrome. The plant's deadly symptoms are caused by atropine's disruption of the parasympathetic nervous system's ability to regulate involuntary activities, such as sweating, breathing, and heart rate. The antidote for belladonna poisoning is physostigmine or pilocarpine, the same as for atropine.

Atropa belladonna is also toxic to many domestic animals, causing narcosis and paralysis. However, cattle and rabbits eat the plant seemingly without suffering harmful effects. In humans, its anticholinergic properties will cause the disruption of cognitive capacities, such as memory and learning.

Uses

Cosmetics: The common name *belladonna* originates from its historic use by women - *Bella Donna* is Italian for *beautiful lady.* Drops prepared from the belladonna plant were used to dilate women's pupils, an effect considered attractive. Belladonna drops act as an antimuscarinic, blocking receptors in the muscles of the eye that constrict pupil size. Belladonna is currently rarely used cosmetically, as it carries the adverse effects of causing minor visual distortions, inability to focus on near objects, and increased heart rate. Prolonged usage was reputed to cause blindness.

Medicinal Uses

Belladonna has been used in herbal medicine for centuries as a pain reliever, muscle relaxer, and anti-inflammatory, and to treat menstrual problems, peptic ulcer disease, histaminic reaction, and motion sickness. At least one 19th-century eclectic medicine journal explained how to prepare a belladona tincture for direct administration to patients.

Belladonna tinctures, decoctions, and powders, as well as alkaloid salt mixtures, are still produced for pharmaceutical use, and these are often standardised at 1037 parts hyoscyamine to 194 parts atropine and 65 parts scopolamine. The alkaloids are compounded with phenobarbital and/or kaolin and pectin for use in various functional gastrointestinal disorders. The tincture, used for identical purposes, remains in most pharmacopoeias, with a similar tincture of *Datura stramonium* having been in the US Pharmacopoeia at least until the late 1930s. The combination of belladonna and opium, in powder, tincture, or alkaloid form, is particularly useful by mouth or as a suppository for diarrhoea and some forms of visceral pain; it can be made by a compounding pharmacist, and may be available as a manufactured fixed combination product in some countries (e.g., B&O Supprettes). A banana-flavoured liquid (most common trade name: Donnagel PG) was available until 31 December 1992 in the United States.

Scopolamine is used as the hydrobromide salt for GI complaints, motion sickness, and to potentiate the analgesic and anxiolytic effects of opioid analgesics. It was formerly used in a painkiller called "twilight sleep" in childbirth.

Atropine sulphate is used as a mydriatic and cycloplegic for eye examinations. It is also used as an antidote to organophosphate and carbamate poisoning, and is loaded in an autoinjector for use in case of a nerve gas attack. Atropinisation (administration of a sufficient dose to block nerve gas effects) results in 100 per cent blockade of the muscarinic acetylcholine receptors and atropine sulphate is the benchmark for measuring the power of anticholinergic drugs.

Hyoscyamine is used as the sulphate or hydrobromide for GI problems and Parkinson's disease. Its side effect profile is intermediate to those of atropine and scopolamine, and can also be used to combat the toxic effects of organophosphates.

Scientific evidence to recommend the use of *A. belladonna* in its natural form for any condition is insuffiicient, although some of its components, in particular *l*-atropine which was purified from belladona in the 1830s, have accepted medical uses. Donnatal is a prescription pharmaceutical, approved in the United States by the FDA, that combines natural belladonna alkaloids in a specific, fixed ratio with phenobarbital to provide peripheral anticholinergic/antispasmodic action and mild sedation. According to its labeling, it is possibly effective for use as adjunctive therapy in the treatment of irritable bowel syndrome (irritable colon, spastic colon, mucous colitis) and acute enterocolitis.

Belladonna preparations are used in homeopathy as treatments for various conditions, although no scientific evidence supports their efficacy. Clinically and in research trials, the most common preparation is diluted to the 30C level in homeopathic notation. This level of dilution does not contain any of the original plant, although preparations with lesser dilutions which statistically contain trace amounts of the plant are advertised for sale.

Recreational Drug

Atropa belladonna and related plants, such as jimson weed (*Datura stramonium*), have occasionally been used as recreational drugs because of the vivid hallucinations and delirium they produce. However, these hallucinations are most commonly described as very unpleasant, and recreational use is considered extremely dangerous because of the high risk of unintentional fatal overdose. In addition, the central nervous system effects of atropine include memory disruption, which may lead to severe confusion.

Poison

The tropane alkaloids of *A. belladonna* were used as poisons, and early humans made poisonous arrows from the plant. In Ancient Rome, it was used as a poison by Agrippina the Younger, wife of Emperor Claudius on advice of Locusta, a lady specialized in poisons, and Livia, who is rumored to have used it to kill her husband Emperor Augustus.

Macbeth of Scotland, when he was still one of the lieutenants of King Duncan I of Scotland, used it during a truce to poison the troops of the invading Harold Harefoot, King of England, to the point that the English troops were unable to stand their ground and had to retreat to their ships.

Folklore

In the past, witches were believed to have used a mixture of belladonna, opium poppy, and other plants, typically poisonous (such as monkshood and poison hemlock) in flying ointment, which they applied to help them fly to gatherings with other witches.

Carlo Ginzburg and others have argued that flying ointments were preparations meant to encourage hallucinatory dreaming; a possible explanation for the inclusion of belladonna and opium poppy in flying ointments concerns the known antagonism between tropane alkaloids of belladonna (specifically scopolamine) and opiate alkaloids in the opium poppy, *Papaver somniferum* (specifically morphine), which produces a dream-like waking state.

This antagonism was known in folk medicine, discussed in eclectic (botanical) medicine formularies, and posited as the explanation of how flying ointments might have actually worked in contemporary writing on witchcraft.

The antagonism between opiates and tropanes is the original basis of the Twilight Sleep that was provided to Queen Victoria to deaden pain as well as consciousness during childbirth, and which was later modified so isolated alkaloids were used instead of plant materials. The belladonna herb was also notable for its unpredictable effects from toxicity.

Baptisia

Baptisia (false indigo, wild indigo) is a genus of about 35 species in the family Fabaceae (legumes). They are flowering herbaceous perennials with pea-like flowers, followed by pods (sometimes inflated). They are native to woodland and grassland in eastern and southern North America. The species most commonly found in cultivation is *B. australis*.

Baptisia species are used as food plants by the larvae of some Lepidoptera species including *Schinia jaguarina*.

List of Species

- *Baptisia alba* (L.) Vent.
- *Baptisia albescens* Small
- *Baptisia arachnifera* W.H. Duncan, two counties of southeastern Georgia (US State)

- *Baptisia australis* (L.) R.Br.
- *Baptisia bicolour* Greenman & Larisey
- *Baptisia bracteata* Elliott
- *Baptisia bushii* Small
- *Baptisia calycosa* Canby
- *Baptisia cinerea* (Raf.) Fernald & B.G.Schub.
- *Baptisia confusa* Pollard & Ball
- *Baptisia cuneata* Small
- *Baptisia deamii* Larisey
- *Baptisia elliptica* Small
- *Baptisia fragilis* Larisey
- *Baptisia fulva* Larisey
- *Baptisia gibbesii* Small
- *Baptisia hirsuta* Small
- *Baptisia hugeri* Small
- *Baptisia intercalata* Larisey
- *Baptisia intermedia* Larisey
- *Baptisia lactea* (Raf.) Thieret
- *Baptisia laevicaulis* (Canby) Small
- *Baptisia lanceolata* (Walter) Elliott, Georgia, Alabama, Florida, and South Carolina
- *Baptisia lecontei* Torr. & A.Gray
- *Baptisia leucantha*
- *Baptisia leucophaea* Nutt.
- *Baptisia macilenta* Small ex Larisey
- *Baptisia megacarpa* Torrey & A.Gray
- *Baptisia minor*
- *Baptisia nuculifera* Greene
- *Baptisia nuttalliana* Small
- *Baptisia oxyphylla* Greene
- *Baptisia pendula* Larisey
- *Baptisia perfoliata* (L.) R.Br. in W.T.Aiton
- *Baptisia pinetorum* Larisey
- *Baptisia psammophila* Larisey

- *Baptisia riparia* Larisey
- *Baptisia saligna* Greene
- *Baptisia simplicifolia* Croom
- *Baptisia sphaerocarpa* Nutt.
- *Baptisia stricta* Larisey
- *Baptisia sulphurea* Englem.
- *Baptisia texana* Pollard & Ball
- *Baptisia tinctoria* (L.) R.Br. in W.T.Aiton
- *Baptisia vespertina* Small ex Rydb.
- *Baptisia viridis* Larisey
- *Baptisia* × *intermedia* Larisey
- *Baptisia* × *variicolour* Kosnik, Diggs, Redshaw & Lipscomb

Berteroa Incana

Berteroa incana (Hoary Alyssum) is a plant in the mustard family, Brassicaceae.

Berteroa incana grows from a deep taproot and is an annual, biennial or short-lived perennial herbaceous plant with alternate, simple gray-green leaves. The foliage is rough with star shaped hairs. Young plants have a basal rosette of leaves that give rise to upright stems with simple, alternate leaves. The leaves are longer than wide. The flowers are white, numerous and arranged into racemes, though they appear to be clustered at the ends of the stems. The flowers have four petals that are deeply notched.

Brassica

Brassica (pron.: /ÈÆbræshÆkYÆ/ *brás-si-ca*) is a genus of plants in the mustard family (Brassicaceae). The members of the genus are collectively known as cruciferous vegetables, cabbages, or mustards. Crops from this genus are sometimes called *cole crops*, which is derived from the Latin *caulis*, meaning *stem or cabbage.*

Common types of brassica used for food include cabbage, cauliflower, broccoli, Brussel sprouts, and some types of seeds. The genus is known for containing many important agricultural and horticultural crops. It also includes a number of weeds, both wild taxa and escapees from cultivation. It includes over 30 wild species and hybrids, and numerous additional cultivars and hybrids of cultivated origin. Most are annuals or biennials, but some are small shrubs. Due to their agricultural importance, *Brassica* plants have been the subject of much scientific interest. Six particularly important species (*Brassica carinata*, *B. juncea*, *B. oleracea*, *B. napus*, *B. nigra* and *B. rapa*) are derived by combining the chromosomes from three earlier species, as described by the Triangle of U theory.

The genus is native in the wild in western Europe, the Mediterranean and temperate regions of Asia. In addition to the cultivated species, which are grown worldwide, many of the wild species grow as weeds, especially in North America, South America, and Australia. A dislike for cabbage, broccoli et.al. can be due to the Brassica species containing a chemical similar to phenylthiocarbamide (PTC), a chemical which is either bitter or tasteless depending on one's genetic makeup.

Uses

Food: Almost all parts of some species or other have been developed for food, including the root (rutabaga, turnips), stems (kohlrabi), leaves (cabbage, kale), flowers (cauliflower, broccoli, Brussel sprouts), and seeds (many, including mustard seed, and oil-producing rapeseed). Some forms with white or purple foliage or flowerheads are also sometimes grown for ornament.

Brassica species are sometimes used as food plants by the larvae of a number of Lepidoptera species.

Medicine

Brassica vegetables are highly regarded for their nutritional value. They provide high amounts of vitamin C and soluble fibre and contain multiple nutrients with potent anticancer properties: 3,3'-diindolylmethane, sulforaphane and selenium. Boiling reduces the level of anticancer compounds, but steaming, microwaving, and stir frying do not result in significant loss. Steaming the vegetable for three to four minutes is recommended to maximize sulforaphane.

Brassica vegetables are rich in indole-3-carbinol, a chemical which boosts DNA repair in cells and appears to block the growth of cancer cells. They are also a good source of carotenoids, with broccoli having especially high levels. Researchers at the University of California at Berkeley have recently discovered that 3,3'-diindolylmethane in *Brassica* vegetables is a potent modulator of the innate immune response system with potent antiviral, antibacterial and anticancer activity; however, it also is an antiandrogen. These vegetables also contain goitrogens, which suppress thyroid function. This can induce hypothyroidism and goiter.

Species

There is some disagreement among botanists on the classification and status of *Brassica* species and subspecies. The following is an abbreviated list, with an emphasis on economically important species.

- *B. carinata*: Abyssinian mustard or Abyssinian cabbage, used to produce biodiesel
- *B. elongata*: elongated mustard
- *B. fruticulosa*: Mediterranean cabbage
- *B. juncea*: Indian mustard, brown and leaf mustards, Sarepta mustard
- *B. napus*: rapeseed, canola, rutabaga (swede/Swedish turnip/ swede turnip)
- *B. narinosa*: broadbeaked mustard
- *B. nigra*: black mustard
- *B. oleracea*: kale, cabbage, broccoli, cauliflower, *kai-lan*, Brussel sprouts, kohlrabi
- *B. perviridis*: tender green, mustard spinach
- *B. rapa* (syn *B. campestris*): Chinese cabbage, turnip, rapini, *komatsuna*
- *B. rupestris*: brown mustard
- *B. septiceps*: seventop turnip
- *B. tournefortii*: Asian mustard

Other Species Formerly Placed in Brassica

- *B. kaber* (wild mustard or charlock)
- *B. alba* or *B. hirta* (white or yellow mustard)
- *B. geniculata* (hoary mustard)

Genome Sequencing and Genetics

Bayer Cropscience (in collaboration with BGI-Shenzhen, China, Keygene N.V., the Netherlands and the University of Queensland, Australia) announced it had sequenced the entire genome of rapeseed/ canola (*Brassica napus*) and its constituent genomes present in *B. rapa* and *B. oleracea* in 2009.

The *B. rapa* genome is currently being sequenced by the Multinational Brassica Genome Project. This also represents the A genome component of the amphidiploid crop species *B. napus* and *B. juncea*.

Buxus Sempervirens

Buxus sempervirens (common box, European box, or boxwood), is a species of flowering plant in the genus *Buxus*, native to western and southern Europe, northwest Africa, and southwest Asia, from southern England south to northern Morocco, and east through the northern Mediterranean region to Turkey. *Buxus colchica* of western Caucasus and *B. hyrcana* of northern Iran and eastern Caucasus are commonly treated as synonyms of *B. sempervirens.*

Description

Buxus sempervirens is an evergreen shrub or small tree growing to 1–9 m (3 ft 3 in–29 ft 6 in) tall, with a trunk up to 20 centimetres (7.9 in) in diametre (exceptionally to 10 m tall and 45 cm diametre). Arranged in opposite pairs along the stems, the leaves are green to yellow-green, oval, 1.5–3 cm long, and 0.5–1.3 cm broad. The hermaphrodite flowers are inconspicuous, greenish-yellow, with no petals, and are insect pollinated; the fruit is a three-lobed capsule containing 3-6 seeds.

Distribution and Habitat

The species typically grows on soils derived from chalk, limestone, usually as an understorey in forests of larger trees, most commonly associated with *Fagus sylvatica* forests, but also sometimes in open dry montane scrub, particularly in the Mediterranean region. Box Hill, Surrey is named after its notable box population, which comprises the largest area of native box woodland in England.

The species is locally naturalised in parts of North America.

Cultivation

In Britain, three burials of the Roman era featured coffins lined with sprays of the evergreen box, a practice unattested elsewhere in Europe. Box remains a very popular ornamental plant in gardens, being particularly valued for topiary and hedges because of its small leaves, evergreen nature, tolerance of close shearing, and scented foliage. The scent is not to everyone's liking: the herbalist John Gerard found it "evill and lothsome" and at Hampton Court Palace Queen Anne had box hedging grubbed up because the odor was offensive, Daniel Defoe tells.

Several cultivars have been selected, including 'Argenteo-variegata' and 'Marginata' with variegated foliage; such "gilded box" received a first notice in John Parkinson's *Paradisi in Sole Paradisus Terrestris*

(1629). 'Vardar Valley', a slow-growing particularly hardy semi-dwarf cultivar, was selected in 1935 by the American botanist Edward Anderson in the upper Vardar valley and sent to the Arnold Arboretum for evaluation.

The following varieties and cultivars have gained the Royal Horticultural Society's Award of Garden Merit:-

- *B. sempervirens*
- *B. sempervirens* 'Elegantissima'
- *B. sempervirens* 'Latifolia Maculata'
- *B. sempervirens* 'Suffruticosa'

Timber

Slow growth of box renders the wood ("boxwood") very hard (possibly the hardest in Europe) and heavy, and free of grain produced by growth rings, making it ideal for cabinet-making, the crafting of clarinets, engraving, marquetry, woodturning, tool handles, mallet heads and as a substitute for ivory. The noted English engraver Thomas Bewick pioneered the use of boxwood blocks for engraving.

Celastrus Scandens

Celastrus scandens, commonly called American Bittersweet or Bittersweet, is a species of staff vines that blooms mostly in June and is commonly found on rich, well-drained soils of woodlands. It has a sturdy perennial vine that may have twining, woody stems that are 30 feet (9.1 m) or longer and an inch or more thick at the base.

The stems are yellowish-green to brown and wind around other vegetation, sometimes killing saplings by restricting further growth. It has tiny, scentless flowers at the tips of the branches. It has colourful, orange fruits that are the size of a pea. These fruits are poisonous to humans when ingested internally, but are favourites of birds. *C. scandens* roots were used by Native Americans and pioneers to induce vomiting, to treat venereal disease, and to treat symptoms of tuberculosis.

C. scandens is native to central and eastern North America. It was given the name *Bittersweet* by European colonists in the 18th century because the fruits resembled the appearance of the fruits of Eurasian Nightshade (*Solanum dulcamara*), which was also called *Bittersweet*. Today, *American Bittersweet* is the accepted common name of *C. scandens* in large part to distinguish it from an invasive relative, *C. orbiculatus* (Oriental Bittersweet), from Asia.

Centaurea

Centaurea is a genus of between 350 and 600 species of herbaceous thistle-like flowering plants in the family Asteraceae. Members of the genus are found only north of the equator, mostly in the Eastern Hemisphere; the Middle East and surrounding regions are particularly species-rich. Common names for this genus are centaury, centory, starthistles, knapweeds, centaureas and the more ambiguous "bluets"; a vernacular name used for these plants in parts of England is "loggerheads". The *Plectocephalus* group – possibly a distinct genus – is known as basketflowers. "Cornflowers" is used for a few species, but that term more often specifically means *C. cyanus* (sometimes also called "Basket Flower"). The common name "centaury" is sometimes used, although this also refers to the unrelated plant genus *Centaurium*.

Description and Ecology

Knapweeds are robust weedy plants. Their leaves, spiny in some species, are usually deeply divided into elongated lobes at least in the plants' lower part, becoming entire towards the top. The "flowers" (actually pseudanthium inflorescences) are diverse in colour, ranging from intense blues, reds and yellows to any mixture of these and lighter shades towards white. Often, the disk flowers are much darker or lighter than the ray flowers, which also differ in morphology and are sterile. Each pseudanthium sits atop a cup- or basket-like cluster of scaly bracts, hence the name "basketflowers". Many species, in particular those inhabiting more arid regions, have a long and strong taproot.

Certain knapweeds have a tendency to dominate large stretches of landscape together with a few other plants, typically one or two grasses and as many other large herbs. The Common Knapweed (*C. nigra*) for example is plentiful in the mesotrophic grasslands of England and nearby regions. It is most prominently found in pastures or meadows dominated by Cock's-foot (*Dactylis glomerata*) as well as either of Crested Dog's-tail (*Cynosurus cristatus*) and False Oat-grass (*Arrhenatherum elatius*). It is also often found in mesotrophic grassland on rendzinas and similar calcareous soils in association with Glaucous Sedge (*Carex flacca*), Sheep's Fescue (*Festuca ovina*), and either Tor-grass (*Brachypodium pinnatum*) and Rough Hawkbit (*Leontodon hispidus*), or Upright Brome (*Bromus erectus*). In these grasslands, Greater Knapweed (*C. scabiosa*) is found much more rarely by comparison, often in association with Red Festuce (*F. rubra*) in addition to Cock's-foot, False Oat-grass.

Due to their habit of dominating ecosystems under good conditions, many *Centaurea* species can become invasive weeds in regions where they are not native. In parts of North America, Diffuse Knapweed (*C. diffusa*), Spotted Knapweed (*C. maculosa*) and Yellow Starthistle (*C. solstitialis*) cause severe problems in agriculture due to their uncontrolled spread. The seeds are typically transported by human traffic, in particular the tires of all-terrain vehicles. The two knapweeds are harmful mainly because they are strongly allelopathic, producing powerful toxins in their roots that stunt the growth of plants around them not adapted to this. Yellow Starthistle, meanwhile, is inedible to most livestock due to its spines and apparently outright poisonous to horses and other equines. However, efficient methods of biological control by insect pests of these weeds have been developed; the knapweeds can also exploited to their detriment by targeted grazing. Controlled burning may also be used, though the timing is important to avoid the plants having seeded already, and neither allowing sufficient time for them to regrow from the rootstock.

Yet other species of *Centaurea* – mostly ones that occur between Italy and the Caucasus – are endemics of a single island or valley, and some of these are endangered. The Akamas Centaurea (*Centaurea akamantis*) of Cyprus is almost extinct, while the western Caucasus endemics *C. leptophylla* and *C. straminicephala* are at least very rare and *C. hedgei* and *C. pecho* from the same region are certainly not abundant either. The last four species would be adversely affected by the proposed Yusufeli Dam, which might actually destroy enough habitat to push the two rarer ones over the brink of extinction.

Centaurea are copious nectar producers, especially on high-lime soils. The high nectar yield of the genus makes it very attractive to insects such as butterflies – including the endangered Karner Blue (*Lycaeides melissa samuelis*) which visits introduced Spotted Knapweed – and day-flying moths – typically Zygaenidae, such as *Zygaena loti* or the Six-spot Burnet (*Z. filipendulae*). The larvae of some other Lepidoptera species use *Centaurea* species as food plants; see List of Lepidoptera that feed on Centaurea. Several of these are used in biological control of invasive knapweeds and starthistles.

Larvae of several true weevils (Curculionidae) of the subfamily Lixinae also feed on *Centaurea*. Some genera – such as *Larinus* whose larval food is flowerheads – have many species especially adapted to particular knapweeds or starthistle and are used in biological control too. These include the Yellow Starthistle Flower Weevil (*L. curtus*) for

Yellow Starthistle, Lesser Knapweed Flower Weevil (*L. minutus*) for Diffuse Knapweed and Blunt Knapweed Flower Weevil (*L. obtusus*) for Spotted Knapweed. Broad-nosed Seedhead Weevil (*Bangasternus fausti*) larvae eat Diffuse, Spotted and Squarrose Knapweed (*C. virgata* ssp. *squarrosa*), while those of the Yellow Starthistle Bud Weevil (*B. orientalis*) do not seem to live on anything other than Yellow Starthistle and occasionally Purple Starthistle (*C. calcitrapa*). But perhaps most efficient in destroying developing Yellow Starthistle seedheads is the larva of the Yellow Starthistle Hairy Weevil (*Eustenopus villosus*). Knapweed Root Weevil (*Cyphocleonus achates*) larvae bore into the roots of Spotted and to a lesser extentely Diffuse Knapweed, sometimes killing off the entire plant.

Also used in biological control are Tephritidae (peacock flies) whose larvae feed on *Centaurea*. Knapweed Peacock Fly (*Chaetorellia acrolophi*) larvae eat Spotted Knapweed and some other species. The Yellow Starthistle Peacock Fly (*C. australis*) has an initial generation each year which often uses Cornflower (*C. cyanus*) as larval food; later generations switch to Yellow Starthistle. The flies are generally considered less efficient in destroying the growing seedheads than the weevils, but may be superior under certain conditions; employing flies and weevils in combination is expensive and does not noticeably increase their effect.

Use by Humans

Though the overall effect of the entire genus on humans is probably detrimental, this is merely due to the considerable damage done by a handful of species when these are allowed to spread uncontrolled. On a per-species basis, most centaureas are inoffensive and at least pleasant-looking plants. As noted above, many species of *Centaurea* are popular food for insects that may otherwise attack crops; particular in small-scale farming it is often advisable to allow non-invasive species (like Cornflower, *C. cyanus*, in Europe) to grow around fields. In such a beetle bank, they will draw pests away from crops, and at the same time by their robust structure provide shelter for invertebrate and small vertebrate predators that will keep overall pest populations low.

The abundant nectar produced by *Centaurea* flowers attracts many pollinators. This is another reason for the success of the highly invasive species, but as it holds true for the entire genus it is not decisive. Yet it can be exploited to the farmer's advantage in combination with biological control of these weeds: In particular the Yellow Starthistle (*C. solstitialis*) as well as Spotted Knapweed (*C. maculosa*) are major honey plants for beekeepers. Monofloral honey from these plants is light and slightly tangy, and one of the finest honeys produced in the USA – due to its better availability, it is even fraudulently relabeled and sold as the scarce and expensive Sourwood honey of the Appalachian Mountains. Placing beehives near stands of *Centaurea* will cause increased pollination.

As most seedheads fail however when biocontrol pests have established themselves, the plants will bloom ever more abundantly in an attempt to replace the destroyed seedheads, to the point where they exhaust their resources in providing food for the pests (seeds), bees (pollen) and humans (honey). Output of allelopathic compounds is also liable to be reduced under such conditions – the plant has to compromise between allocating energy to reproduction and defense.

This renders the weeds more likely to be suppressed by native vegetation or crops in the following years, especially if properly timed controlled burning and/or targeted grazing by suitable livestock are also employed. While Yellow Starthistle and perhaps other species are toxic to equines, some other livestock may eat the non-spiny knapweeds with relish. In Europe, Common Knapweed (*C. nigra*) and Globe Knapweed (*C. macrocephala*) are locally important pollen sources for honeybees in mid-late summer.

8-Hydroxyquinoline has been identified as a main allelopathic compound produced by Diffuse Knapweed (*C. diffusa*); native North American plants are typically sensitive to it, while those of Eastern Europe and Asia Minor usually have coevolved with the knapweed and are little harmed if at all, aided by native microorganisms that break down or even feed on the abundantly secreted compound. Thus, 8-hydroxyquinoline is potentially useful to control American plants that have become invasive weeds in the Diffuse Knapweed's native range.

Arctiin, found in *C. imperialis*, has shown anticancer activity in laboratory studies. The roots of the long-lost *C. foliosa*, an endemic of Hatay Province (Turkey), are used in folk medicine, and other species are presumably too. A South Italian variety of the Purple Starthistle (*C. calcitrapa*) is traditionally consumed by ethnic Albanians (Arbëreshë people) in the Vulture area (southern Italy); e.g. in the Arbëreshë communities in Lucania the young whorls of *C. calcitrapa* are boiled and fried in mixtures with other weedy non-cultivated greens.

According to research by the Michael Heinrich group at the Centre for Pharmacognosy and Phytotherapy (School of Pharmacy, University of London) "the antioxidant activity [...] of the young whorls of *Centaurea calcitrapa*, both in the DPPH and in the lipid peroxidation inhibition assays, [is] very interesting and [the] species should be investigated phytochemically and biochemically focusing on these properties". Extracts from *C. calcitrapa* were furthermore found to have significant xanthine oxidase (XO)-inhibiting activity.

Spotted Knapweed as well as other species are rich in cnicin, a bitter compound found mainly in the leaves and often used to flavor the digestif amaro. In western Crete, Greece a local variety of *C. calcitrapa* called *gourounaki* (ãïõñïõíÜêé "little pig") also has its leaves eaten boiled by the locals. In the same island an endemic local species, *C. idaea* called *katsoula* (êáôóïýëá), *tsita* (ôóßôá) or *aspragatha* (áóðñáãêÜèá), has its leaves eaten boiled by the locals too.

Some species are cultivated as ornamental plants in gardens. As regards other aspects of popular culture, Cornflower (*C. cyanus*) is the floral emblem of Östergötland province (Sweden) – where is it called *blåklint*, literally "blue mountain" – and of Päijänne Tavastia region in Finland, where it is known as *ruiskaunokki* ("rye-beaks") or *ruiskukka* ("rye-flower").

It is also the national flower of Estonia – its local name *rukkilill* meaning "rye-lily" –, Belarus where it is called *va³oška* (Belarusian: âàëîøêà), and one of those of Germany where it is called *Kornblume* ("cornflower"). The origin of the name "caltrop" for the ancient low-tech area denial weapon is probably in some way connected with *C. calcitrapa* and its spiny seeds. This plant is attested to by the colloquial name "caltrop" at a time when the weapons were still called by their Roman name *tribulus*. Lastly, the colour cornflower blue is named after *C. cyanus*.

Systematics and Taxonomy

As namesake member of the subtribe Centaureinae of tribe Cynareae, the knapweeds are probably most closely related to genera such as *Carthamus* (distaff thistles), *Cnicus* (Blessed Thistle), *Crupina* (crupinas) or *Notobasis* (Syrian Thistle), and somewhat less closely to most other thistles.

The monotypic *Cnicus* seems in fact to properly belong in *Centaurea*. Research in the late 20th century shows that *Centaurea* as traditionally defined is polyphyletic. A number of 19th- and 20th-century efforts to reorganize the genus were not successful, and it is not yet clear what the consequences of the recent research will be for classification of this genus and other related genera.

The type species *C. centaurium* stands somewhat apart from the main lineage of knapweeds and thus the taxonomic consequences of a rearrangement might be severe, with hundreds of species needing to be moved to new genera. It has thus been proposed to change the type species to one of the main lineage to avoid this problem. What seems certain however is that the basketflowers – presently treated as a section *Plectocephalus* – will be reinstated as a distinct genus in the near future. The rock-centauries (*Cheirolophus*), formerly usually included in *Centaurea*, are now already treated as separate genus.

Synonyms

- *Acosta* Adans.
- *Aegialophila* Boiss. & Heldr.
- *Calcitrapa* Vaill.
- *Calcitrapoides* Vaill.
- *Chartolepis* Cass.
- *Cheirolepis* Boiss.
- *Cnicus* L., *nom. cons.*
- *Colymbada* Hill
- *Crocodilium* Vaill.
- *Grossheimia*Sosn. & Takht.
- *Hyalea* (DC.) Jaub. & Spach
- *Jacea* Mill.
- *Melanoloma* Cass.
- *Phaeopappus* (DC.) Boiss.
- *Plectocephalus* D. Don
- *Stephanochilus* Maire
- *Tomanthea* DC.
- *Wagenitzia* Dostál
- *Amberboa* sect. *Phaeopappus* DC.
- *Centaurea* sect. *Hyalaea* DC.

Selected Species

Species presently placed in *Centaurea* include:

- *Centaurea adpressa*
- *Centaurea aegyptiaca*
- *Centaurea aeolica*
- *Centaurea aggregata*

- *Centaurea akamantis* – Akamas Centaurea
- *Centaurea alba*
- *Centaurea albonitens* Turrill
- *Centaurea alpestris*
- *Centaurea alpina*
- *Centaurea ambigua*
- *Centaurea amblyolepis*
- *Centaurea americana* – American Basketflower, American Starthistle
- *Centaurea ammocyanus*
- *Centaurea antennata* Dufour
- *Centaurea antiochia* Boiss.
- *Centaurea aplolepa*
    * *Centaurea aplolepa* ssp. *carueliana*
- *Centaurea appendicigera* C.Koch
- *Centaurea argentea*
- *Centaurea ascalonica*
- *Centaurea aspera* L. – Rough Starthistle
- *Centaurea atacamensis* (Reiche) I.M.Johnst.
- *Centaurea atropurpurea*
- *Centaurea ×aurata*
- *Centaurea babylonica* L.
- *Centaurea balsamita*
- *Centaurea behen* L. – *ak behmen* (Turkish)
- *Centaurea bella*
- *Centaurea bieberseinii*
- *Centaurea borjae*
- *Centaurea bovina*
- *Centaurea bracteata*
- *Centaurea brevifimbriata* Hub.-Mor.
- *Centaurea bulbosa*
- *Centaurea busambarensis* Guss.
- *Centaurea cachinalensis*
- *Centaurea calcitrapa* – Purple Starthistle, Red Starthistle, "caltrop"

- *Centaurea calcitrapoides*
- *Centaurea cariensis* Boiss.
- *Centaurea cariensiformis* Hub.-Mor.
- *Centaurea centaurium* L.
- *Centaurea chilensis*
- *Centaurea cineraria* – Velvet Centaurea, "dusty miller"
- *Centaurea clementei*
- *Centaurea collina* L.
- *Centaurea corymbosa*
- *Centaurea crithmifolia*
- *Centaurea crocodylium*
- *Centaurea cyanoides* J.Berggr. & Wahlenb.
- *Centaurea cyanus* – Cornflower, Bachelor's Button, Boutonniere Flower, Hurtsickle, "bluebottle", "basketflower"
- *Centaurea damascena*
- *Centaurea dealbata* – Persian Cornflower, Whitewash Cornflower
- *Centaurea debeauxii* Gren. & Godr.
- *Centaurea depressa* – Low Cornflower
- *Centaurea deusta*
- *Centaurea diffusa* – Diffuse Knapweed, White Knapweed, Tumble Knapweed
- *Centaurea diluta* – North African Knapweed
- *Centaurea drabifolia* Sm.
- *Centaurea dschungarica*
- *Centaurea eriophora*
- *Centaurea eryngioides*
- *Centaurea filiformis*
- *Centaurea floccosa*
- *Centaurea foliosa* Boiss. & Kotschy
- *Centaurea forojuliensis*
- *Centaurea friderici* Vis. – *palagruška zeèina* (Croatian)
- *Centaurea gayana*
- *Centaurea glaberrima* Tausch
- *Centaurea glastifolia*

- *Centaurea grinensis*
- *Centaurea gymnocarpa*
- *Centaurea haradjianii* Wagenitz
- *Centaurea hedgei*
- *Centaurea helenioides* Boiss.
- *Centaurea hermannii* F.Hermann
- *Centaurea horrida* Badarò – *fiordaliso spinoso* (Italian)
- *Centaurea hyalolepis*
- *Centaurea hypoleuca*
- *Centaurea iberica* – Iberian Starthistle, Iberian Knapweed
- *Centaurea idaea* – *katsoula, tsita* (Cretan Greek)
- *Centaurea imperialis* Hausskn. ex Bornm.
- *Centaurea jabukensis*
- *Centaurea jacea* – Brown Knapweed, Brownray Knapweed
- *Centaurea kasakorum*
- *Centaurea kopetaghensis*
- *Centaurea kotschyana* Heuff.
- *Centaurea lanulata*
- *Centaurea leptophylla*
- *Centaurea leucophylla*
- *Centaurea limbata*
- *Centaurea lydia* Boiss.
- *Centaurea macrocephala* Puschk. ex Willd. – Globe Knapweed, Armenian Basketflower
- *Centaurea maculosa* – Spotted Knapweed (might belong in *C. stoebe* ssp. *micranthos*)
- *Centaurea mannagettae*
- *Centaurea margaritalba* Klok.
- *Centaurea marschalliana*
- *Centaurea melitensis* – Maltese Starthistle; *tocalote, tocolote* (California)
- *Centaurea minor*
- *Centaurea moschata* – Sweet Sultan
- *Centaurea ×moncktonii* C.E.Britton – Meadow Knapweed, Protean Knapweed (= *C. ×pratensis* Thuill non Salisb.)

- *Centaurea monocephala*
- *Centaurea montana* – Montane Knapweed, Perennial Cornflower, Mountain Cornflower, Mountain Bluet
- *Centaurea napifolia* L. – *fiordaliso romano* (Italian)
- *Centaurea nervosa* Rchb. ex Steud.
- *Centaurea nigra* – Common Knapweed, Black Knapweed, Lesser Knapweed, Hardheads
- *Centaurea nigrescens* – Tyrol Knapweed, Short-fringed Knapweed, Tyrol Thistle
- *Centaurea nigrifimbria* (C.Koch) Sosn.
- *Centaurea nivea* (Bornm.) Wagenitz
- *Centaurea onopordifolia*
- *Centaurea orientalis* L.
- *Centaurea ornata* Willd.
- *Centaurea ovina*
- *Centaurea pallescens* Delile
- *Centaurea paniculata* L.
- *Centaurea parlatoris*
- *Centaurea pecho*
- *Centaurea phrygia* – Wig Knapweed
- *Centaurea pindicola*
- *Centaurea polypodiifolia*
- *Centaurea ×pratensis* Salisb. (*C. jacea* × *C. nigra*) – Meadow Knapweed
- *Centaurea procurrens*
- *Centaurea ×psammogena* G.Gayer. (*C. diffusa* × *C. stoebe* ssp. *micranthos*)
- *Centaurea pseudocaerulescens*
- *Centaurea pseudophrygia* C.A.Mey.
- *Centaurea pulcherrima* Willd.
- *Centaurea pullata* L.
- *Centaurea pumilio*
- *Centaurea ragusina* L.
- *Centaurea rigida*
- *Centaurea rothrockii* Greenm. – Mexican Basketflower, Rothrock's Basketflower, Rothrock's Knapweed

- *Centaurea ruthenica*
- *Centaurea rutifolia* Sm.
- *Centaurea sadleriana* – Pannonian Knapweed
- *Centaurea salicifolia* Bieb. ex Willd.
- *Centaurea scabiosa* – Greater Knapweed
- *Centaurea scannensis*
- *Centaurea scoparia*
- *Centaurea scopulorum* Boiss. & Heldr.
- *Centaurea seguenzae*
- *Centaurea seridis* L.
- *Centaurea sibirica*
- *Centaurea simplicicaulis*
- *Centaurea sinaica*
- *Centaurea solstitialis* – Yellow Starthistle, Golden Starthistle, Yellow Cockspur, St. Barnaby's Thistle, Barnaby Thistle
- *Centaurea speciosa*
- *Centaurea sphaerocephala* L.
- *Centaurea stenolepis*
- *Centaurea stoebe* L.
    * *Centaurea stoebe* ssp. *micranthos* (Gugler) Hayek
- *Centaurea straminicephala*
- *Centaurea sulphurea* – Sicilian Starthistle
- *Centaurea tauromenitana* Guss.
- *Centaurea tenoreana*
- *Centaurea tommasinii*
- *Centaurea transalpina* Schleich. ex DC.
- *Centaurea tchihatcheffii* — *yanardöner* (Turkish)
- *Centaurea trichocephala* Bieb. ex Willd. – Featherhead Knapweed
- *Centaurea triniifolia*
- *Centaurea triumfettii* All.
- *Centaurea ucriae* Lacaita
- *Centaurea uniflora* Turra
- *Centaurea verbascifolia* Vahl

- *Centaurea verutum* L.
- *Centaurea virgata*
    - * *Centaurea virgata* ssp. *squarrosa* – Squarrose Knapweed
- *Centaurea wiedemanniana* Fisch. & Mey.
- *Centaurea yozgatensis* Wagenitz

Formerly Placed Here

Plant species placed in *Centaurea* in former times include:

- *Acroptilon repens* – Russian Knapweed (as *C. repens*)
- *Cheirolophus crassifolius* – Maltese Rock-centaury (as *C. crassifolia, C. spathulata*)
- *Femeniasia balearica* (as *C. balearica*)
- *Volutaria muricata* (as *C. muricata*)

Chapter 2

Research and Statistics

Plants cannot move to escape their predators, so they must have other means of protecting themselves from herbivorous animals. Some plants have physical defenses such as thorns, but by far the most common protection is chemical. Over millennia, natural selection has produced a complicated and vast array of chemical compounds that deter herbivores. Tannin is a compound that emerged relatively early in the evolutionary history of plants, while more complex molecules such as polyacetylenes are found in younger groups of plants such as the Asterales.

Many of the plant defense compounds arose to defend against consumption by insects, although when livestock or humans consume such plants, they may also experience negative effects, ranging from mild discomfort to death.

Many of these poisonous compounds also have important medicinal benefits. There are so many kinds of plant defenses that there are many unanswered questions about them. Questions include:

(1) which plants have which type of defenses,

(2) which herbivores are the plants defended against,

(3) what are the chemical structures of the compounds that provide defense,

(4) and what are the potential medical uses of these compounds?

This is still an active area of research with important implications for understanding plant evolution, and for medical research.

Below is an extensive, if incomplete, list of plants containing poisonous parts that pose a serious risk of illness, injury, or death to humans or animals. Human fatalities caused by poisonous plants – especially resulting from accidental ingestion – are rare in the USA.

Poisonous Food Plants

Many food plants possess toxic parts, are toxic unless processed, or are toxic at certain stages of their life. Notable examples include:

- Apple (*Malus domestica*). Seeds are mildly poisonous, containing a small amount of amygdalin, a cyanogenic glycoside. The quantity contained is usually not enough to be dangerous to humans, but it is possible to ingest enough seeds to provide a fatal dose.
- Cassava (*Manihot esculenta*) Roots and leaves contain two cyanogenic glucosides, linamarin and lotaustralin. These are decomposed by linamarase, a naturally occurring enzyme in cassava, liberating hydrogen cyanide. Cassava varieties are often categorized as either sweet or bitter, respectively signifying the absence or presence of toxic levels of cyanogenic glucosides. The 'sweet' cultivars can produce as little as 20 milligrams of cyanide per kilogram of fresh roots, whereas bitter ones may produce more than 50 times as much (1 g/kg). Cassavas grown during drought are especially high in these toxins. A dose of 40 mg of pure cassava cyanogenic glucoside is sufficient to kill a cow.

  It can also cause severe calcific pancreatitis in humans, leading to chronic pancreatitis. Processing (soaking, cooking, fermentation, etc.) of cassava root is necessary to remove the toxins and avoid getting sick. "Chronic, low-level cyanide exposure is associated with the development of goiter and with tropical ataxic neuropathy, a nerve-damaging disorder that renders a person unsteady and uncoordinated. Severe cyanide poisoning, particularly during famines, is associated with outbreaks of a debilitating, irreversible paralytic disorder called konzo and, in some cases, death. The incidence of konzo and tropical ataxic neuropathy can be as high as 3 percent in some areas." For some smaller-rooted sweet varieties, cooking is sufficient to eliminate all toxicity. The cyanide is carried away in the processing water and the amounts produced in domestic consumption are too small to have environmental impact. The larger-rooted, bitter varieties used for production of flour or starch must be processed to remove the cyanogenic glucosides. Industrial production of cassava flour, even at the cottage level, may generate enough cyanide and cyanogenic glycosides in the effluents to have a severe environmental impact.

- Cherry (*Prunus cerasus*), as well as other *Prunus* species such as peach (*Prunus persica*), plum (*Prunus domestica*), almond (*Prunus dulcis*), and apricot (*Prunus armeniaca*). Leaves and seeds contain cyanogenic glycosides.
- Indian pea (*Lathyrus sativus*). A legume grown in Asia and East Africa as an insurance crop for use during famines. Like other grain legumes, *L. sativus* produces a high-protein seed. The seeds contain variable amounts of â-N-Oxalyl-L-á,â-diaminopropionic acid or ODAP, a neurotoxic amino acid. ODAP causes wasting and paralysis if eaten over a long period, and is considered as the cause of the disease neurolathyrism, a neurodegenerative disease that causes paralysis of the lower body and emaciation of gluteal muscle (buttocks). The disease has been seen to occur after famines in Europe (France, Spain, Germany), North Africa and South Asia, and is still prevalent in Eritrea, Ethiopia and parts of Afghanistan when *Lathyrus* seed is the exclusive or main source of nutrients for extended periods.
- Kidney bean or common bean (*Phaseolus vulgaris*). The toxic compound phytohaemagglutinin, a lectin, is present in many varieties of common bean but is especially concentrated in red kidney beans. The lectin has a number of effects on cell metabolism; it induces mitosis, and affects the cell membrane in regard to transport and permeability to proteins. It agglutinates most mammalian red blood cell types. The primary symptoms of phytohaemagglutinin poisoning are nausea, vomiting, and diarrhea. Onset is from 1 to 3 hours after consumption of improperly prepared beans, and symptoms typically resolve within a few hours. Consumption of as few as four or five raw kidney beans may be sufficient to trigger symptoms. Phytohaemagglutinin can be deactivated by cooking beans at 100 °C (212 °F) for ten minutes.

  However, for dry beans the U.S. Food and Drug Administration (FDA) also recommends an initial soak of at least 5 hours in water; the soaking water should be discarded. The ten minutes at 100 °C (212 °F) is required to degrade the toxin, and is much shorter than the hours required to fully cook the beans themselves. However, lower cooking temperatures may have the paradoxical effect of potentiating the toxic effect of haemagglutinin. Beans cooked at 80 °C (176 °F) are reported to be up five times as toxic as raw beans. Outbreaks of poisoning have been associated with the use of slow cookers, the low cooking temperatures of which may be unable to degrade the toxin.

- Nutmeg (*Myristica fragrans*). Contains myristicin. Myristicin is a naturally occurring insecticide and acaricide with possible neurotoxic effects on neuroblastoma cells. It has psychoactive properties at doses much higher than used in cooking. Raw nutmeg produces anticholinergic-like symptoms, attributed to myristicin and elemicin. The intoxicating effects of myristicin can lead to a physical state somewhere between waking and dreaming; euphoria is reported and nausea is often experienced. Users also report bloodshot eyes and memory disturbances. Myristicin is also known to induce hallucinogenic effects, such as visual distortions. Nutmeg intoxication has an extremely long time before peak is reached, sometimes taking up to seven hours, and effects can be felt for 24 hours, with lingering effects lasting up to 72 hours.
- Lima bean or butter bean (*Phaseolus lunatus*). Raw beans contain dangerous amounts of linamarin, a cyanogenic glucoside.
- Lupin. Some varieties have edible seeds. Sweet Lupins have less, and Bitter Lupins have more of the toxic alkaloids lupinine and sparteine.
- Onions and garlic. Onions and garlic (genus *Allium*) contain thiosulphate, which in high doses is toxic to dogs, cats and some other livestock.
- Potato (*Solanum tuberosum*). Potatoes contain toxic compounds known as glycoalkaloids, of which the most prevalent are solanine and chaconine. Solanine is also found in other members of the Solanaceae plant family, which includes *Atropa belladonna* ("deadly nightshade") and *Hyoscyamus niger* ("henbane"). The concentration of glycoalkaloid in wild potatoes suffices to produce toxic effects in humans. The toxin affects the nervous system, causing headaches, diarrhea and intense digestive disturbances, cramps, weakness and confusion, and in severe cases coma and death. Poisoning from cultivated potatoes occurs very rarely however, as the toxic compounds in the potato plant are, in general, concentrated in the green portions of the plant and in the fruits, and cultivated potato

varieties contain lower toxin levels. Cooking at high temperatures (over 170 °C or 340 °F) also partly destroys the toxin. However, exposure to light, physical damage, and age increase glycoalkaloid content within the tuber, the highest concentrations occurring just underneath the skin. Tubers which are exposed to light turn green from chlorophyll synthesis, thus giving a visual clue as to areas of the tuber that may have become more toxic; however, this does not provide a definitive guide, as greening and glycoalkaloid accumulation can occur independently of each other.

Some varieties of potato contain greater glycoalkaloid concentrations than others; breeders developing new varieties test for this, and sometimes have to discard an otherwise promising cultivar. Breeders try to keep solanine levels below 200 mg/kg (200 ppmw). However, when these commercial varieties turn green, even they can approach concentrations of solanine of 1000 mg/kg (1000 ppmw). The U.S. National Toxicology Programme suggests that the average American consume at most 12.5 mg/day of solanine from potatoes (the toxic dose is actually several times this, depending on body weight). Douglas L. Holt, the State Extension Specialist for Food Safety at the University of Missouri, notes that no reported cases of potato-source solanine poisoning have occurred in the U.S. in the last 50 years, and most cases involved eating green potatoes or drinking potato-leaf tea.

- Rhubarb (*Rheum rhaponticum*). The leaf stalks (petioles) are edible, but the leaves themselves contain notable quantities of oxalic acid, which is a nephrotoxic and corrosive acid that is present in many plants. Symptoms of poisoning include kidney disorders, convulsions and coma. Rarely fatal. The LD_{50} (median lethal dose) for pure oxalic acid in rats is about 375 mg/kg body weight, or about 25 grams for a 65 kg (~140 lb) human. While the oxalic acid content of rhubarb leaves can vary, a typical value is about 0.5%, so a rather unlikely 5 kg of the extremely sour leaves would have to be consumed to reach an LD_{50} of oxalic acid. Cooking the leaves with soda can make them more poisonous by producing soluble oxalates.

  However, the leaves are believed to also contain an additional, unidentified toxin, which might be an anthraquinone glycoside (also known as senna glycosides). In the edible leaf stalks (petioles), the amount of oxalic acid is much lower, only about 2-

2.5% of the total acidity which is dominated by malic acid. This means that even the raw stalks may not be hazardous (though they are generally thought to be in the US). However the tart taste of raw stalks is so strong as to be unpalatable to many.

- Tomato (*Solanum lycopersicum*). Like many other nightshades, tomato leaves and stems contain solanine that is toxic if ingested, causing digestive upset and nervous excitement. Use of tomato leaves as a tea (tisane) has been responsible for at least one death. Leaves, stems, and green unripe fruit of the tomato plant also contain small amounts of the poisonous alkaloid tomatine, although levels are generally too small to be dangerous. Ripe tomatoes do not contain any detectable tomatine. Tomato plants can be toxic to dogs if they eat large amounts of the fruit, or chew plant material.

Other Poisonous Plants

- *Abrus precatorius* (known commonly as jequirity, crab's eye, rosary pea, '*John Crow'* bead, precatory bean, Indian licorice, akar saga, giddee giddee, jumbie bead, ruti, and weather plant). The attractive seeds (usually about the size of a ladybug, glossy red with one black dot) contain abrin, which is related to ricin, and very potent. Symptoms of poisoning include nausea, vomiting, convulsions, liver failure, and death, usually after several days. Ingesting a single seed can kill an adult human. The seeds have been used as beads in jewellery, which is dangerous; inhaled dust is toxic and pinpricks can be fatal. The seeds are unfortunately attractive to children.
- *Aconitum* (Several species, commonly called aconite, wolfsbane and monkshood) All parts are poisonous. The poison is an alkaloid called aconitine, which disables nerves, lowers blood pressure, and can stop the heart. Even casual skin contact should be avoided; symptoms include numbness, tingling, and cardiac irregularity. It has been used as poison for bullets (by Germany in WWII), as a bait and arrow poison (ancient Greece), and to poison water supplies (reports from ancient Asia). If ingested, it usually causes burning, tingling, and numbness in the mouth, followed by vomiting and nervous excitement. It is usually a quick-acting poison. Used in the past for killing wolves (hence one of the common names).
- *Actaea pachypoda* (also known as doll's eyes or white baneberry). All parts are poisonous, but especially the berries, the

consumption of which has a sedative effect on cardiac muscle tissue and can cause cardiac arrest.

- Adam and Eve
- *Adenium obesum* (also known as sabi star, kudu or desert-rose). Exudes a highly toxic sap which is used by the Meridian High and Hadza in Tanzania to coat arrow-tips for hunting.
- *Aesculus hippocastanum* (commonly known as horse-chestnut). All parts of the plant are poisonous, causing nausea, muscle twitches, and sometimes paralysis.
- African sumac
- *Agave.* The juice of a number of species causes acute contact dermatitis, with blistering lasting several weeks and recurring itching for several years thereafter.
- *Ageratina altissima* (commonly known as white snakeroot). All parts are poisonous, causing nausea and vomiting. Often fatal. Milk from cattle that have eaten white snakeroot can sicken, or kill, humans (milk sickness).
- *Agrostemma githago* (commonly known as corn cockle). Contains the saponins githagin and agrostemmic acid. All parts of the plant are reported to be poisonous and may produce chronic or acute, potentially fatal poisoning, although it has been used in folk medicine to treat a range of ills, from parasites to cancer . There are no known recent clinical studies of corn cockle which provide a basis for dosage recommendations, however doses higher than 3 g [of seeds] are considered toxic.[2]
- *Akar saga*
- *Amianthium*
- *Angel's Trumpet*
- *Angel Wings*
- *Anticlea*
- *Aquilegia* (also known as columbine). Several species. Seeds and roots contain cardiogenic toxins which cause both severe gastroenteritis and heart palpitations if consumed. The flowers of various species were consumed in moderation by Native Americans as a condiment with other fresh greens, and are reported to be very sweet, and safe if consumed in small quantities. Native Americans also used very small amounts of the root as an effective treatment for ulcers. However, the medical use of this plant is difficult due to its high toxicity; columbine poisonings are easily fatal.

- *Areca catechu* (commonly known as betel nut palm and pinyang). The nut contains an alkaloid related to nicotine which is addictive. It produces a mild high, some stimulation, and lots of red saliva, which cannot be swallowed as it causes nausea. Withdrawal causes headache and sweats. Use is correlated with mouth cancer, and to a lesser extent asthma and heart disease.
- *Arum maculatum* (commonly known as cuckoo-pint, lords and ladies, jack in the pulpit, wake robin, wild arum, devils and angels, cows and bulls, Adam and Eve, bobbins and starch-root). All parts of the plant can produce allergic reactions. The bright red berries contain oxalates of saponins and can cause skin, mouth and throat irritation, resulting in swelling, burning pain, breathing difficulties and stomach upset. One of the most common causes of plant poisoning.
- *Asparagus*. The berries are poisonous.
- *Atropa belladonna* (commonly known as deadly nightshade, belladonna, devil's cherry and dwale, an Anglo-Saxon term meaning stupifying drink). One of the most toxic plants found in the Western hemisphere. All parts of the plant contain tropane alkaloids. The active agents are atropine, hyoscine (scopolamine), and hyoscyamine, which have anticholinergic properties. The symptoms of poisoning include dilated pupils, sensitivity to light, blurred vision, tachycardia, loss of balance, staggering, headache, rash, flushing, dry mouth and throat, slurred speech, urinary retention, constipation, confusion, hallucinations, delirium, and convulsions.

  The root of the plant is generally the most toxic part, though this can vary from one specimen to another. Ingestion of a single leaf of the plant can be fatal to an adult. Casual contact with the leaves can cause skin pustules. The berries pose the greatest danger to children because they look attractive and have a somewhat sweet taste.

  The consumption of two to five berries by children and ten to twenty berries by adults can be lethal. In 2009 a case of *A. belladonna* being mistaken for blueberries, with six berries ingested by an adult woman, was documented to result in severe anticholinergic syndrome. The plant's deadly symptoms are caused by atropine's disruption of the parasympathetic nervous system's ability to regulate involuntary activities such as sweating, breathing, and heart rate. The antidote for atropine poisoning is physostigmine or pilocarpine. *A. belladonna* is also toxic to many domestic animals, causing narcosis and paralysis. However, cattle and rabbits eat the plant seemingly without suffering harmful effects. In humans its anticholinergic properties will cause the disruption of cognitive capacities like memory and learning.

- *Autumn crocus*
- *Azalea*
- *Bittersweet nightshade*
- *Black hellebore*
- *Black locust*
- *Black nightshade*
- *Bleeding heart*
- *Blind-your-eye mangrove*
- *Blister Bush*
- *Bloodroot*
- *Blue-green algae*
- *Bobbins*
- *Bracken*
- *Broom*
- *Brugmansia* (commonly known as angel's trumpet). All parts of the plant contain the tropane alkaloids scopolamine and atropine. Often fatal.
- *Calabar Bean*
- *Caladium* (commonly known as angel wings, elephant ear and heart of Jesus). All parts of the plant are poisonous. Symptoms are generally irritation, pain, and swelling of tissues. If the mouth or tongue swell, breathing may be fatally blocked.
- *Castor oil plant*
- *Cerbera odollam* (commonly known as the suicide tree). The seeds contain cerberin, a potent toxin related to digoxin. The poison blocks the calcium ion channels in heart muscle, causing disruption of the heart beat. This is typically fatal and can result from ingesting a single seed. Cerberin is difficult to detect in autopsies and its taste can be masked with strong spices, such as a curry. It is often used in homicide and suicide in India; Kerala's suicide rate is about three times the Indian average. In 2004, a team led by Yvan Gaillard of the Laboratory of

Analytical Toxicology in La Voulte-sur-Rhône, France, documented more than 500 cases of fatal Cerbera poisoning between 1989 and 1999 in Kerala. They said "To the best of our knowledge, no plant in the world is responsible for as many deaths by suicide as the *odollam* tree.' A related species is *Cerbera tanghin* the seeds of which are known as *tanghin poison nut* and have been used as an 'ordeal poison'.

- *Chelidonium majus* (also known as greater celandine). The whole plant is toxic in moderate doses as it contains a range of isoquinoline alkaloids, but there are claimed to be therapeutic uses when used at the correct dosage. The main alkaloid present in the herb and root is coptisine, with berberine, chelidonine, sanguinarine and chelerythrine also present. Sanguinarine is particularly toxic with an LD_{50} of only 18 mg per kg body weight. The effect of the fresh herb is analgesic, cholagogic, antimicrobial and oncostatic, with action as a central nervous system sedative. In animal tests, *Chelidonium majus* is shown to be cytostatic. Early studies showed that the latex causes contact dermatitis and eye irritation. Stains on skin of the fingers are sometimes reported to cause eye irritation after rubbing the eyes or handling contact lenses. The characteristic latex also contains proteolytic enzymes and the phytocystatin chelidostatin, a cysteine protease inhibitor.
- *Christmas ros*
- *Cicuta* (several species) (commonly known as water hemlock, cowbane, wild carrot, snakeweed, poison parsnip, false parsley, children's bane and death-of-man). The root, when freshly pulled out of the ground, is extremely poisonous and contains the toxin *cicutoxin*, a central nervous system stimulant, resulting in seizures. When dried, the poisonous effect is reduced. The most common species is *C. maculata*; one of the species found in the Western USA, *C. douglasii*, often found in pastures and swamps, has especially thick stems and very large and sturdy flowers which are sometimes harvested for flower displays. This is inadvisable as the sap is also toxic.
- *Cocklebur*
- *Colchicum autumnale* (commonly known as autumn crocus and meadow saffron). The bulbs contain colchicine. Colchicine poisoning has been compared to arsenic poisoning; symptoms start 2 to 5 hours after the toxic dose has been ingested and

include burning in the mouth and throat, fever, vomiting, diarrhea, abdominal pain and kidney failure. These symptoms may set in as many as 24 hours after the exposure. Onset of multiple-system organ failure may occur within 24 to 72 hours. This includes hypovolemic shock due to extreme vascular damage and fluid loss through the GI tract, which may result in death. Additionally, sufferers may experience kidney damage resulting in low urine output and bloody urine; low white blood cell counts (persisting for several days); anemia; muscular weakness; and respiratory failure. Recovery may begin within 6 to 8 days. There is no specific antidote for colchicine, although various treatments do exist. Despite dosing issues concerning its toxicity, colchicine is prescribed in the treatment of gout, familial Mediterranean fever, pericarditis and Behçet's disease. It is also being investigated for its use as an anti-cancer drug.

- *Columbine*
- *Conium maculatum* (commonly known as hemlock, poison hemlock, spotted parsley, spotted cowbane, bad-man's oatmeal, poison snakeweed and beaver poison). All parts of the plant contain the alkaloid coniine which causes stomach pains, vomiting, and progressive paralysis of the central nervous system. Can be fatal; it is the poison that killed Socrates. Not to be confused with hemlock trees (*Tsuga* spp), which, while not edible, are not nearly as toxic as the herbaceous plant *Conium.*
- *Consolida* (commonly known as larkspur). Young plants and seeds are poisonous, causing nausea, muscle twitches, paralysis. Often fatal.
- *Convallaria majalis* (commonly known as lily of the valley). Contains 38 different cardiac glycosides.
- *Coriaria myrtifolia* (commonly known as redoul). A mediterranean plant containing the toxin coriamyrtin, ingestion of which produces digestive, neurological and respiratory problems. The poisonous fruits superficially resemble blackberries and may mistakenly be eaten as such. Can be fatal in children.
- *Corn cockle*
- *Corn lily*
- *Cowbane*

- *Cows and bulls*
- *Crab's eye*
- *Cuckoo-pint*
- Cyanobacteria A phylum of bacteria, commonly known as blue-green algae. Many different species, including *Anacystis cynea* and *Anabaena circinalis*. Produce several different toxins known collectively as cyanotoxins. These can include neurotoxins, hepatotoxins, endotoxins and cytotoxins. Potentially hazardous particularly to marine animals, but also to humans.
- *Cytisus scoparius* (commonly known as broom or common broom). Contains toxic alkaloids that depress the heart and nervous system. The alkaloid sparteine is a class 1a antiarrhythmic agent; a sodium channel blocker. It is not FDA approved for human use as an antiarrhythmic agent, and it is not included in the Vaughn Williams classification of antiarrhythmic drugs.
- *Daffodil*
- *Daphne*. The berries (either red or yellow) are poisonous, causing burns to mouth and digestive tract, followed by coma. Often fatal.
- *Darnel*
- *Datura* Contains the alkaloids scopolamine and atropine. Datura has been used as a hallucinogenic drug by the native peoples of the Americas and others. Incorrect dosage can lead to death.
- *Datura stramonium* (commonly known as jimson weed, thorn apple, stinkweed and Jamestown weed). All parts of the plant are poisonous, causing abnormal thirst, vision distortions, delirium, incoherence, coma. Often fatal. A significant grazing livestock poison in North America.
- *Deadly nightshade*
- *Deathcamas* – various genera in the Melanthieae have species whose common name includes "deathcamas", including *Amianthium*, *Anticlea*, *Stenanthium*, *Toxicoscordion* and *Zigadenus*. All parts of the plants are toxic, due to the presence of alkaloids. Grazing animals, such as sheep and cattle, may be affected and human fatalities have occurred.
- *Delphinium* (also known as larkspur). Contains the alkaloid delsoline. Young plants and seeds are poisonous, causing nausea, muscle twitches, paralysis, often fatal.

- *Dendrocnide moroides* (also known as stinging tree and gympie gympie). Capable of inflicting a painful sting when touched. The stinging may last for several days and is exacerbated by touching, rubbing, and cold. Can be fatal.
- *Devils and angels*
- *Dicentra cucullaria* (also known as bleeding heart and Dutchman's breeches). Leaves and roots are poisonous and cause convulsions and other nervous symptoms.
- *Dichapetalum cymosum* (also known as gifblaar). Well known as a livestock poison in South Africa; this plant contains the metabolic poison fluoroacetic acid.
- *Dieffenbachia* (commonly known as *dumbcane*'). All parts are poisonous, causing intense burning, irritation, and immobility of the tongue, mouth, and throat. Swelling can be severe enough to block breathing, leading to death.
- *Digitalis purpurea* (commonly known as foxglove). The leaves, seeds, and flowers are poisonous, containing cardiac or other steroid glycosides. These cause irregular heartbeat, general digestive upset, and confusion. Can be fatal.
- *Doll's eyes*
- *Dumbcane*
- *Dutchman's breeches*
- *Elder / Elderberry*
- *Euonymus europaeus* (commonly known as spindle, European spindle or spindle tree). The fruit is poisonous, containing amongst other substances, the alkaloids theobromine and caffeine, as well as an extremely bitter terpene. Poisonings are more common in young children, who are enticed by the brightly coloured fruits. Ingestion can result in liver and kidney damage and even death. There are many other species of *Euonymus*, many of which are also poisonous.
- *Excoecaria agallocha* (commonly known as milky mangrove, blind-your-eye mangrove and river poison tree). Contact with latex can cause skin irritation and blistering; eye contact can cause temporary blindness.
- *False acacia*
- *False hellebore*
- *Foxglove*

- *Frangipani*
- *Gelsemium sempervirens* (commonly known as yellow jessamine). All parts are poisonous, causing nausea and vomiting. Often fatal. It is possible to become ill from ingesting honey made from jessamine nectar.
- *Giant hogweed*
- *Giddee giddee*
- *Gifblaar*
- *Greater celandine*
- *Gympie gympie*
- *Heart of Jesus*
- *Hedera helix* (or common ivy) The leaves and berries are poisonous, causing stomach pains, laboured breathing, possible coma.
- *Helleborus niger* (also known as Christmas rose) Contains protoanemonin, or ranunculin, which has an acrid taste and can cause burning of the eyes, mouth and throat, oral ulceration, gastroenteritis and hematemesis.
- *Hemlock*
- *Hemlock water-dropwort*
- *Henbane*
- *Heracleum mantegazzianum* (also known as giant hogweed). The sap is phototoxic, causing phytophotodermatitis (severe skin inflammations) when affected skin is exposed to sunlight or to UV-rays. Initially the skin colours red and starts itching. Then blisters form as reaction continues over 48 hours. They form black or purplish scars, which can last several years. Hospitalisation may become necessary. Presence of minute amounts of sap in the eyes can lead to temporary or even permanent blindness.
- *Hippomane mancinella* (commonly known as manchineel). All parts of this tree, including the fruit, contain toxic phorbol esters typical of the Euphorbiaceae plant family. Specifically the tree contains 12-deoxy-5-hydroxyphorbol-6gamma, 7alpha-oxide, hippomanins, mancinellin, sapogenin, phloracetophenone-2, 4-dimethylether is present in the leaves, while the fruits possess physostigmine. Contact with the milky white latex produces strong allergic dermatitis. Standing beneath the tree during rain will cause blistering of the skin from even slight contact

with this liquid (even a small drop of rain with the milky substance in it will cause the skin to blister). Burning tree parts may cause blindness if the smoke reaches the eyes. The fruit can also be fatal if eaten. Many trees carry a warning sign, while others have been marked with a red "X" on the trunk to indicate danger. In the French Antilles the trees are often marked with a painted red band a few feet above the ground. The Caribs used the latex of this tree to poison their arrows and would tie captives to the trunk of the tree, ensuring a slow and painful death. A poultice of arrowroot (*Maranta arundinacea*) was used by the Arawaks and Taíno as an antidote against such arrow poisons. The Caribs were also known to poison the water supply of their enemies with the leaves. Spanish explorer Juan Ponce de León was struck by an arrow that had been poisoned with manchineel sap during battle with the Calusa in Florida, dying shortly thereafter.

- *Horse chestnut*
- *Holly (European)*
- *Hyacinth*
- *Hyacinthus orientalis* (commonly known as hyacinth). The bulbs are poisonous, causing nausea, vomiting, gasping, convulsions, and possibly death. Even handling the bulbs can cause skin irritation.
- *Hyoscyamus niger* (commonly known as henbane). Seeds and foliage contain hyoscyamine, scopolamine and other tropane alkaloids. Can produce dilated pupils, hallucinations, increased heart rate, convulsions, vomiting, hypertension and ataxia.
- *Ilex aquifolium* (commonly known as European holly). The berries cause gastroenteritis, resulting in nausea, vomiting and diarrhoea.
- *Indian licorice*
- Ivy (Common)
- *Jack in the pulpit*
- *Jacobaea vulgaris* (commonly known as ragwort). Contains many different alkaloids, including jacobine, jaconine, jacozine, otosenine, retrorsine, seneciphylline, senecionine, and senkirkine. Poisonous to livestock and hence of concern to people who keep horses and cattle. Horses do not normally eat fresh ragwort due to its bitter taste, however it loses this taste when dried, and become dangerous in hay. The result, if sufficient

quantity is consumed, can be irreversible cirrhosis of the liver. Signs that a horse has been poisoned include yellow mucus membranes, depression, and lack of coordination. The danger is that the toxin can have a cumulative effect; the alkaloid does not actually accumulate in the liver but a breakdown product can damage DNA and progressively kills cells. *Jacobaea vulgaris* is also theoretically poisonous to humans, although poisoning is unlikely as it is distasteful and not used as a food. However some sensitive individuals can suffer from an allergic skin reaction after handling the plant because, like many members of the compositae family, it contains sesquiterpine lactones (which are different from the pyrrolizidine alkaloids which are responsible for the toxic effects), which can cause compositae dermatitis.

- *Jamestown weed*
- *Jequirity*
- *Jerusalem cherry*
- *Jimson weed*
- *'John Crow' Bead*
- *Jumbie bead*
- *Kalanchoe delagoensis* (commonly known as mother of millions) Contains bufadienolide cardiac glycosides which can cause cardiac poisoning, particularly in grazing animals. During 1997, 125 head of cattle died after eating mother-of-millions on a travelling stock reserve near Moree, NSW.
- *Kalmia latifolia* (commonly known as mountain laurel). Contains andromedotoxin and arbutin. The green parts of the plant, flowers, twigs, and pollen are all toxic, and symptoms of toxicity begin to appear about 6 hours following ingestion. Poisoning produces anorexia, repeated swallowing, profuse salivation, depression, uncoordination, vomiting, frequent defecation, watering of the eyes, irregular or difficulty breathing, weakness, cardiac distress, convulsions, coma, and eventually death. Autopsy will show gastrointestinal irritation and hemorrhage.
- *Laburnum*. All parts of the plant and especially the seeds are poisonous and can be lethal if consumed in excess. The main toxin is cytisine, a nicotinic receptor agonist. Symptoms of poisoning may include intense sleepiness, vomiting, excitement, staggering, convulsive movements, slight frothing at the mouth,

unequally dilated pupils, coma and death. In some cases, diarrhea is very severe and at times the convulsions are markedly tetanic.

- Larkspur
- *Ligustrum* (several species, commonly known as privet). Berries and leaves are poisonous. Berries contain syringin, which causes digestive disturbances, nervous symptoms. Can be fatal. Privet is one of several plants which are poisonous to horses. Privet pollen is known to cause asthma and eczema in sufferers. It is banned from sale or cultivation in New Zealand due to the effects of its pollen on asthma sufferers.
- *Lilium* (commonly known as lily). Most have an unknown water-soluble toxin found in all parts of the plant. Extremely poisonous, yet attractive, to cats, causing acute renal failure; 2 petals can kill.
- *Lily*
- *Lily of the valley*
- *Lolium temulentum* (commonly called darnel or poison ryegrass). The seeds and seed heads of this common garden weed may contain the alkaloids temuline and loliine. Some experts also point to the fungus ergot or fungi of the genus endoconidium, both of which grow on the seed heads of rye grasses, as an additional source of toxicity.
- *Lords and ladies* – see *Arum maculatum.*
- *Madiera winter cherry*
- *Manchineel tree*
- *Mango tree* – Mango peel and sap contains urushiol, the chemical in poison ivy and poison sumac that can cause urushiol-induced contact dermatitis in susceptible people. Cross-reactions between mango contact allergens and urushiol have been observed. Those with a history of poison ivy or poison oak contact dermatitis may be most at risk for such an allergic reaction. Urushiol is also present in mango leaves and stems. During mango's primary ripening season, it is the most common source of plant dermatitis in Hawaii.
- *Mayapple*
- *Meadow saffron*
- *Menispermum* (commonly known as moonseed). The fruits and seeds are poisonous, causing nausea and vomiting. Often fatal.

- *Milky mangrove*
- *Monkshood*
- *Moonseed*
- *Mother of millions*
- *Mountain laurel*
- *Narcissus* (commonly known as daffodil). Various species and garden cultivars. The bulbs are poisonous and cause nausea, vomiting, and diarrhea. Can be fatal. Stems also cause headaches, vomiting, and blurred vision.
- *Nerium oleander* (commonly known as oleander). All parts are toxic, but especially the leaves and woody stems. Contains nerioside, oleandroside, saponins and cardiac glycosides. Causes severe digestive upset, heart trouble and contact dermatitis. The smoke of burning oleander can cause reactions in the lungs, and can be fatal.
- Oak
- *Oenanthe crocata* (commonly known as hemlock water dropwort). Contains oenanthotoxin. The leaves may be eaten safely by livestock, but the stems and especially the carbohydrate-rich roots are much more poisonous. Animals familiar with eating the leaves may eat the roots when these are exposed during ditch clearance – one root is sufficient to kill a cow, and human fatalities are also known in these circumstances.

  Scientists at the University of Eastern Piedmont in Italy claimed to have identified this as the plant responsible for producing the sardonic grin, and it is the most-likely candidate for the "sardonic herb," which was a neurotoxic plant used for the ritual killing of elderly people in Phoenician Sardinia. When these people were unable to support themselves, they were intoxicated with this herb and then dropped from a high rock or beaten to death. Criminals were also executed in this way.
- *Oleander*
- *Ongaonga*
- *Passiflora caerulea* (also known as the blue passion flower or the common passion flower). The leaves contain cyanogenic glycoside, which breaks down into cyanide.
- *Passion flower (blue or common)*

- *Peucedanum galbanum* (commonly known as blister bush). All parts are poisonous, and contact causes painful blistering that is intensified with exposure to sunlight.
- *Physostigma venenosum* (commonly known as calabar bean and also as ordeal beans due to their former use in trials by ordeal). The toxin in the seeds is the parasympathomimetic alkaloid physostigmine, a reversible cholinesterase inhibitor. Symptoms of poisoning include copious saliva, nausea, vomiting, diarrhea, anorexia, dizziness, headache, stomach pain, sweating, dyspepsia and seizures., and can lead to cholinergic syndrome or "SLUDGE syndrome". Medicinal uses of physostigmine include the treatment of myasthenia gravis, glaucoma, Alzheimer's disease and delayed gastric emptying.
- *Plumeria* (commonly known as frangipani). Contact with the milky latex may irritate eyes and skin.
- *Phytolacca* (commonly known as pokeweed). Leaves, berries and roots contain *phytolaccatoxin* and *phytolaccigenin.* Toxin in young leaves is reduced with repeated boiling and draining. Ingestion of poisonous parts of the plant may cause severe stomach cramping, persistent diarrhoea, nausea, vomiting (sometimes bloody), slow and difficult breathing, weakness, spasms, hypertension, severe convulsions, and death.
- *Podophyllum peltatum* (commonly known as mayapple). Green portions of the plant, unripe fruit, and especially the rhizome contain the non-alkaloid toxin podophyllotoxin, which causes diarrhea, severe digestive upset.
- *Poison hemlock*
- *Poison ivy*
- *Poison oak*
- *Poison parsnip*
- *Poison sumac*
- *Poison ryegrass*
- *Pokeweed*
- *Precatory bean*
- *Privet*
- *Pteridium aquilinum* (commonly known as bracken). Carcinogenic to humans and animals such as mice, rats, horses and cattle when ingested. The carcinogenic compound is ptaquiloside or PTQ, which can leach from the plant into the

water supply, which may explain an increase in the incidence of gastric and oesophageal cancers in humans in bracken-rich areas.

- *Quercus* (several species, commonly known as oak)). The leaves and acorns of oak species are poisonous in large amounts to humans and livestock, including cattle, horses, sheep and goats, but not pigs. Poisoning is caused by the toxin tannic acid, which causes gastroenteritis, heart trouble, contact dermatitis and kidney damage. Symptoms of poisoning include lack of appetite, depression, constipation, diarrhea (which may contain blood), blood in urine, and colic. Rarely fatal however, and in fact after proper processing acorns are consumed as a staple in many parts of the world.
- Ragwort
- Redoul
- *Rhododendron* (certain species commonly known as Azaleas). All parts are poisonous and cause nausea, vomiting, depression, breathing difficulties, coma. Rarely fatal.
- *Rhus lancia* (commonly known as African sumac). Closely related to poison ivy, all parts of this tree contain low levels of a highly irritating oil with urushiol. Skin reactions can include blisters and rashes. It spreads readily to clothes and back again, and has a very long life. Infections can follow scratching. As urushiol is not a poison but an allergen, it will not affect certain people. The smoke of burning *Rhus lancia* can cause reactions in the lungs, and can be fatal.
- *Ricinus communis* (commonly known as castor oil plant or Palma Christi). The seeds contain ricin, an extremely toxic water-soluble protein. Also present are ricinine, an alkaloid, and an irritant oil. According to the 2007 edition of the Guinness Book of World Records, this plant is the most poisonous in the world. Castor oil, long used as a laxative, muscle rub, and in cosmetics, is made from the seeds, but the ricin is removed during processing. The lethal dose in adults is considered to be 4 to 8 seeds, but reports of actual poisoning are relatively rare. If ingested, symptoms may be delayed by up to 36 hours but commonly begin within 2–4 hours. These include a burning sensation in mouth and throat, abdominal pain, purging and bloody diarrhea. Within several days there is severe dehydration, a drop in blood pressure and a decrease in urine. Unless treated,

death can be expected to occur within 3–5 days; if victims have not succumbed after this time, they often recover. In 1978, ricin was used to assassinate Georgi Markov, a Bulgarian dissident. He was stabbed with the point of an umbrella while waiting at a bus stop near Waterloo Station in London. After his death a perforated metallic pellet was found embedded in his leg; this had presumably contained the ricin toxin. Toxicity varies among animal species: 4 seeds will kill a rabbit, 5 a sheep, 6 an ox or horse, 7 a pig, and 11 a dog. Poisoning occurs when animals ingest broken seeds or break the seed by chewing; intact seeds may pass through the digestive tract without releasing the toxin. Ducks have shown substantial resistance to the seeds: it takes an average of 80 to kill them.

- *River poison tree*
- *Robinia pseudoacacia* (also known as black locust and false acacia). Pods are toxic.
- Rosary pea
- *Sambucus* (commonly known as elder or elderberry). The roots are considered poisonous and cause nausea and digestive upset.
- *Sanguinaria canadensis* (commonly known as bloodroot). The rhizome contains morphine-like benzylisoquinoline alkaloids, primarily the toxin sanguinarine. Sanguinarine kills animal cells by blocking the action of Na^+/K^+-ATPase transmembrane proteins. As a result, applying *S. canadensis* to the skin may destroy tissue and lead to the formation of a large scab, called an eschar. Although applying escharotic agents, including *S. canadensis*, to the skin is sometimes suggested as a home treatment for skin cancer, these attempts can be severely disfiguring, as well as unsuccessful. Case reports have shown that in such instances tumor has recurred and/or metastasized. The United States Food and Drug Administration (FDA) has approved the inclusion of sanguinarine in toothpastes as an antibacterial or anti-plaque agent, although it is believed that this use may cause leukoplakia, a premalignant oral lesion. The safe level of sanguinarine in such products is subject to regulation and debate. *S. canadensis* extracts have also been promoted by some supplement companies as a treatment or cure for cancer, but the FDA has listed some of these products among its "187 Fake Cancer 'Cures' Consumers Should Avoid". Bloodroot is a popular red natural dye used by Native American artists, especially among southeastern rivercane basketmakers.

However in spite of supposed curative properties and historical use by Native Americans as an emetic, due to its toxicity internal use is not advisable (sanguinarine has an LD_{50} of only 18 mg per kg body weight).

- *Solanum dulcamara* (commonly known as bittersweet nightshade). All parts are poisonous, containing solanine and causing fatigue, paralysis, convulsions, and diarrhea. Rarely fatal.
- *Solanum nigrum* (commonly known as black nightshade). All parts of the plant except the *ripe* fruit contain the toxic glycoalkaloid solanine. Solanine poisoning is primarily displayed by gastrointestinal and neurological disorders. Symptoms include nausea, diarrhea, vomiting, stomach cramps, burning of the throat, cardiac dysrhythmia, headache and dizziness. In more severe cases, hallucinations, loss of sensation, paralysis, fever, jaundice, dilated pupils and hypothermia can result. In large quantities, solanine poisoning can be fatal.
- *Solanum pseudocapsicum* (commonly known as Jerusalem cherry, Madiera winter cherry and winter cherry). All parts, especially the berries, are poisonous, causing nausea and vomiting. It is occasionally fatal, especially to children.
- *Sosnowsky's Hogweed.* Plant has toxic sap and causes skin inflammation on contact.
- *Spindle tree or spindle*
- *Starch-root*
- *Stenanthium*
- *Stinging tree*
- *Stinkweed*
- *Strychnine tree*
- *Strychnos nux-vomica* (commonly known as the strychnine tree). The seeds usually contain about 1.5% strychnine, an extremely bitter and deadly alkaloid. This substance throws a human into intense muscle convulsions and usually kills within three hours. The bark of the tree may also contain brucine, another dangerous chemical.
- *Suicide tree*
- *Taxus baccata* (commonly known as English yew', common yew and graveyard tree). Nearly all parts contain toxic taxanes (except the red, fleshy, and slightly sweet aril surrounding the

toxic seeds). The seeds themselves are particularly toxic if chewed. Several people have committed suicide by ingesting leaves and seeds, including Catuvolcus, king of a tribe in what is now Belgium.

- *Thorn apple*
- *Toxicodendron* Several species, including *Toxicodendron radicans* (commonly known as poison ivy), *Toxicodendron diversilobum* (commonly known as poison-oak), and *Toxicodendron vernix* (commonly known as poison sumac). All parts of these plants contain a highly irritating oil with urushiol. Skin reactions can include blisters and rashes. It spreads readily to clothes and back again, and has a very long life. Infections can follow scratching. Despite the common names, urushiol is actually not a poison but an allergen, and because of this it will not affect certain people. The smoke of burning poison ivy can cause reactions in the lungs, and can be fatal.
- *Toxicoscordion*
- *Urtica ferox* (commonly known as ongaonga). Even the lightest touch can result in a painful sting that lasts several days.
- *Veratrum* (commonly known as false hellebore and corn lily). Several species, containing highly toxic steroidal alkaloids (e.g. veratridine) that activate sodium ion channels and cause rapid cardiac failure and death if ingested. All parts of the plant are poisonous, with the root and rhizomes being the most toxic. Symptoms typically occur between 30 minutes and 4 hours after ingestion and include nausea and vomiting, abdominal pain, numbness, headache, sweating, muscle weakness, bradycardia, hypotension, cardiac arrhythmia, and seizures. Treatment for poisoning includes gastrointestinal decontamination with activated charcoal followed by supportive care including fluid replacement, antiemetics for persistent nausea and vomiting, atropine for treatment of bradycardia, and vasopressors for the treatment of hypotension. Native Americans used the juice pressed from the roots to poison arrows before combat. The dried powdered root of this plant was also used as an insecticide. The plants' teratogenic properties and ability to induce severe birth defects were well known to Native Americans, although they also used minute amounts of the winter-harvested root (combined with *Salvia dorii* to potentiate its effects and reduce the toxicity of the herb) to treat cancerous tumors. The toxic steroidal alkaloids are produced only when the plants are in

active growth, so herbalists and Native Americans who used this plant for medicinal purposes harvested the roots during the winter months when the levels of toxic constituents were at their lowest. The roots of *V. nigrum* and *V. schindleri* have been used in Chinese herbalism (where plants of this genus are known as "li lu" (Ü...†). Li lu is used internally as a powerful emetic of last resort, and topically to kill external parasites, treat tinea and scabies, and stop itching.

However some herbalists refuse to prescribe li lu internally, citing the extreme difficulty in preparing a safe and effective dosage, and that death has occurred at a dosage of 0.6 grams. During the 1930s *Veratrum* extracts were investigated in the treatment of high blood pressure in humans. However patients often suffered side effects due to the narrow therapeutic index of these products. Due to its toxicity, the use of *Veratrum* as a treatment for high blood pressure in humans was discontinued.

- *Wake robin*
- *Water hemlock*
- *White baneberry*
- *White snakeroot*
- *Wild arum*
- *Winter cherry*
- *Wolfsbane*
- *Xanthium* (commonly known as cocklebur). Several species. The Common Cocklebur (*X. strumarium*), a native of North America, can be poisonous to livestock, including horses, cattle, and sheep. Some domestic animals will avoid consuming the plant if other forage is present, but less discriminating animals, such as pigs, will consume the plants and then sicken and die.

  The seedlings and seeds are the most toxic parts of the plants. Symptoms usually occur within a few hours, producing unsteadiness and weakness, depression, nausea and vomiting, twisting of the neck muscles, rapid and weak pulse, difficulty breathing, and eventually death. *Xanthium* has also been used for its medicinal properties and for making yellow dye, as indicated by its name (Greek *xanthos* = 'yellow').
- *Yellow jessamine*
- *Yew*

- *Zantedeschia* (several species, also known as Lily of the Nile and Calla lily). Contain calcium oxalate. All parts of the plant are toxic, producing irritation and swelling of the mouth and throat, acute vomiting and diarrhoea. Can be fatal.
- *Zigadenus*

Plants Poisonous to Equines

Many plants are poisonous to equines; the species vary depending on location, climate and grazing conditions. In many cases, entire genera are poisonous to equines and include many species spread over several continents. Plants can cause reactions ranging from laminitis (found in horses bedded on shavings from black walnut trees), anemia, kidney disease and kidney failure (from eating the wilted leaves of red maples), to cyanide poisoning (from the ingestion of plant matter from members of the *Prunus* genus) and other symptoms. Members of the *Prunus* genus have also been theorized to be at fault for mare reproductive loss syndrome. Some plants, including yew, are deadly and extremely fast-acting. Several plants, including nightshade, become more toxic as they wilt and die, posing a danger to horses eating dried hay or plant matter blown into their pastures.

The risk of animals becoming ill during the fall is increased, as many plants slow their growth in preparation for winter, and equines begin to browse on the remaining plants. Many toxic plants are unpalatable, so animals will avoid them where possible. However, this is not always the case; locoweeds, for example, are addictive and once a horse has eaten them, it will continue to eat them whenever possible, and can never be exposed to them again. When a toxic plant is ingested, it can be difficult to diagnose, because exposure over time can cause symptoms to occur after the animal is no longer exposed to the plant. Toxins are often metabolized before the symptoms become obvious, making it hard or impossible to test for them. Hungry or thirsty horses are more likely to eat poisonous plants, as are those pastured on overgrazed lands. Animals with mineral deficiencies due to poor diets will sometimes seek out poisonous plants. Poisonous plants are more of a danger to livestock after wildfires, as they often regrow more quickly.

Abrus Precatorius

Abrus precatorius, known commonly as jequirity, Crab's eye, rosary pea, John Crow Bead, precatory bean, Indian licorice, Akar Saga, gidee gidee or Jumbie bead in Trinidad & Tobago, is a slender, perennial climber that twines around trees, shrubs, and hedges. It is a legume with long, pinnate-leafleted leaves.

The plant is best known for its seeds, which are used as beads and in percussion instruments, and which are toxic due to the presence of abrin. The plant is native to Indonesia and grows in tropical and subtropical areas of the world where it has been introduced. It has a tendency to become weedy and invasive where it has been introduced.

Toxin

The toxin abrin is a dimer consisting of two protein subunits, termed A and B. The B chain facilitates abrin's entry into a cell by bonding to certain transport proteins on cell membranes, which then transport the toxin into the cell. Once inside the cell, the A chain prevents protein synthesis by inactivating the 26S subunit of the ribosome. One molecule of abrin will inactivate up to 1,500 ribosomes per second.

Symptoms are identical to those of ricin, except abrin is more toxic by almost two orders of magnitude; the fatal dose of abrin is approximately 75 times smaller than the fatal dose of ricin. Abrin can kill with a circulating amount of less than 3 micrograms. Abrin has an estimated human fatal dose of 0.1-1 ug/kg. Ingesting the intact seeds typically results in no clinical findings, as they pass through the gastrointestinal tract due to their hard shell.

Abrus precatorius, called *kudri mani* in Tamil and *guru ginja* in Telugu, has been used in Siddha medicine for centuries. The Tamil Siddhars knew about the toxic effects in plants and suggested various methods which is called "suththi seythal" or purification. This is done by boiling the seeds in milk and then drying them. The protein is denatured when subjected to high temperatures which removes it toxicity.

In March 2012 a recall was issued for bracelets made using Jequirity Beans sold by the Eden Project and other outlets in the UK.

This plant is also poisonous to horses.

Uses

The seeds of *Abrus precatorius* are much valued in native jewellery for their bright coloration. Most beans are black and red, suggesting a ladybug, though other colours are available. Jewellery-making with jequirity seeds is dangerous, and there have been cases of death by a finger-prick while boring the seeds for beadwork.

In Trinidad in the West Indies the brightly coloured seeds are strung into bracelets and worn around the wrist or ankle to ward off jumbies or evil spirits and "mal-yeux" - the evil eye. The Tamils use *Abrus* seeds of different colours. The red variety with black eye is the most common, but there are black, white and green varieties as well.

The seeds of *Abrus precatorius* are very consistent in weight. Formerly Indians used these seeds to weigh gold using a measure called a *Ratti*, where 8 Ratti = 1 Masha; 12 Masha = 1 Tola (11.6 Grams).

In Northern India The seeds are usually used in worship of Sri Salagrama Representing Srimati Radharani. In the Gaudiya Vaisnava History, it is known that Sri Caitanya Mahaprabhu Would Compare the Gunja Seeds of that to Srimati Radharani's feet.

Traditional Medicine

In Siddha medicine, the white variety is used to prepare oil that is claimed to be an aphrodisiac. A tea is made from the leaves and used to treat fevers, coughs and colds. Seeds are poisonous and therefore are used after mitigation. The plant is also used in Ayurveda.

Laboratory Study of Extracts

A variety of pharmacological effects have been observed in rodents, but have not been demonstrated clinically in humans, including:

- An ethanolic extract of *Abrus precatorius* was found to have antioxidant, anti-inflammatory and analgesic potential in rodents.
- A methanolic extract of *A. precatorius* seeds causes reversible alterations in the estrous cycle pattern and completely blocked ovulation in Sprague-Dawley rats.
- The methanolic extract produces dose-dependent bronchodilator activity in a guinea pig model.

Names

Abrus precatorius has different names in various Indian languages.

- *Assamese :* Latumoni
- *Sanskrit :* Gunja
- *Hindi :* Ratti; Gaungchi; Gunchi; Gunja
- *Bengali :* Kunch; Koonch
- *Gujarati :* Gumchi; Chanothi
- *Kannada :* Gulaganji (—Á²—,œ¿)
- *Kashmiri :* Shangir

- *Malayalam :* Kunni; Kunnikkuru
- *Persian :* Gunchi; Chashami-Khurosa
- *Punjabi :* Mulati
- *Tamil :* Gundumani; Kunthamani
- *Telugu :* Gurivinda (A0?5?&)
- *Marwari:*Chirmi
- *Oriya:*Kaincha
- *öArabic :* Úíä ÇáÚÝÑíÊ
- *Nepali :* Ratigedi
- *Hebrew :* Avrus (àáøåñ)

Acer Rubrum

Acer rubrum (Red Maple, also known as Swamp, Water or Soft Maple), is one of the most common and widespread deciduous trees of eastern North America. The U.S. Forest service recognizes it as the most common variety of tree in America. The red maple ranges from the Lake of the Woods on the border between Ontario and Minnesota, east to Newfoundland, south to near Miami, Florida, and southwest to east Texas. Many of its features, especially its leaves, are quite variable in form. At maturity it often attains a height of around 15 m (49 ft). It is aptly named as its flowers, petioles, twigs and seeds are all red to varying degrees. Among these features, however, it is best known for its brilliant deep scarlet foliage in autumn.

Over most of its range, red maple is adaptable to a very wide range of site conditions, perhaps more so than any other tree in eastern North America. It can be found growing in swamps, on poor dry soils, and most anywhere in between. It grows well from sea level to about 900 m (3,000 ft). Due to its attractive fall foliage and pleasing form, it is often used as a shade tree for landscapes. It is used commercially on a small scale for maple syrup production as well as for its medium to high quality lumber. It is also the State Tree of Rhode Island.

Description

Though *A. rubrum* is usually easy to identify, it is highly changeable in morphological characteristics. It is a medium to large sized tree, reaching heights of 18 to 27 metres (59 to 89 ft) and exceptionally over 35 metres (115 feet). The leaves are usually 9 to 11 centimetres (3.5 to 4.3 in) long on a full grown tree. The trunk diametre can range from 46 to 76 cm (18 to 30 in), depending on the growing conditions. Its spread is about 12 m (39 ft). A 10-year-old sapling will stand about 6 m (20 ft) tall. In forests, the bark will remain free of branches until some distance up the tree. Individuals grown in the open are shorter and thicker with a more rounded crown. Generally speaking, however, the crown is irregularly ovoid with ascending whip-like curved shoots. The bark is a pale grey and smooth when the individual is young. As the tree grows the bark becomes darker and cracks into slightly raised long plates. The largest known living red maple is located near Armada, Michigan, at a height of 38.1 m (125 ft) and a bole circumference, at breast height, of 4.95 m (16.2 ft).

The leaves of the red maple offer the easiest way to distinguish it from its relatives. As with nearly all maple trees, they are deciduous and arranged oppositely on the twig. They are typically 5–10 cm (2.0–3.9 in) long and wide with 3-5 palmate lobes with a serrated margin. The sinuses are typically narrow, but the leaves can exhibit considerable variation. When 5 lobes are present, the three at the terminal end are larger than the other two near the base. In contrast, the leaves of the related silver maple, *A. saccharinum*, are much more deeply lobed, more sharply toothed and characteristically have 5 lobes. The upper side of *A. rubrum*'s leaf is light green and the underside is whitish and can be either glaucous or hairy. The leaf stalks are usually red and are up to 10 cm (3.9 in) long. Furthermore, the leaves turn a brilliant red in autumn.

The twigs of the red maple are reddish in colour and somewhat shiny with small lenticels. Dwarf shoots are present on many branches. The buds are usually blunt and greenish to reddish in colour, generally with several loose scales. The lateral buds are slightly stalked, and in addition there may be collateral buds present as well. The buds form in fall and winter and are often visible from a distance due to their reddish tint. The leaf scars on the twig are V-shaped and contain 3 bundle scars.

The flowers are generally unisexual, with male and female flowers appearing in separate sessile clusters, though they are sometimes also bisexual. They appear in spring from April to May, usually coming before the leaves. The tree itself is considered Polygamodioecious, meaning some individuals are male, some female, and some monoecious. The flowers are red with 5 small petals and a 5-lobed calyx borne in hanging clusters, usually at the twig tips. They are lineal to oblong in shape and are pubescent. The pistillate flowers have one pistil formed from two fused carpels with a glabrous superior ovary and two long styles that protrude beyond the perianth. The staminate flowers contain between 4 and 12 stamens, often with 8.

The fruit is a 15 to 25 millimetre (.5 to .75 inch) long double samara with somewhat divergent wings at an angle of 50 to 60 degrees. They are borne on long slender stems and are variable in colour from light brown to reddish. They ripen from April through early June, before even the leaf development is altogether complete. After they reach maturity, the seeds are dispersed for a 1 to 2 week period from April through July.

Distribution and Habitat

A. rubrum is one of the most abundant and widespread trees in eastern North America. It can be found from the south of Newfoundland, Nova Scotia and southern Quebec to the south west of Ontario, extreme southeastern Manitoba and northern Minnesota; south to Wisconsin, Illinois, Missouri, eastern Oklahoma, and eastern Texas in its western range; and east to Florida. It has the largest continuous range along the North American Atlantic Coast of any tree that occurs in Florida. In total it ranges 2,600 km (1,600 mi) from north to south. The species is native to all regions of the United States east of the 95th meridian west, with only three exceptions, namely the Prairie Peninsula of the Midwest, the coastal prairie in southern Louisiana and southeastern Texas and the swamp prairie of the Florida Everglades. In several other locations, the tree is absent from large areas but still present in a few specific habitats. An example is the Bluegrass region of Kentucky, where it is not found in the dominant open plains, but is present along streams. Here the red maple is not present in the bottom land forests of the Grain Belt, despite the fact it is common in similar habitats and species associations both to the north and south of this area.

The tree's range ends where the "40 °C ("40 °F) mean minimum isotherm begins, namely in southeastern Canada. On the other hand, the western range is limited by the much drier climate of the Great Plains. Nonetheless, it has the widest tolerance to climatic conditions of all the North American species of maple. The absence of red maple in the Prairie Peninsula is due to the species intolerance of fire.

A. rubrum does very well in a wide range of soil types, with varying textures, moisture, pH, and elevation, probably more so than any other forest tree in North America. It grows on glaciated as well as nonglaciated soils derived from the following rocks: granite, gneiss, schist, sandstone, shale, slate, conglomerate, quartzite, and limestone. Chlorosis can occur on very alkaline soils, though otherwise its pH tolerance is quite high. As concerns levels of moisture, the red maple grows everywhere from dry ridges and southwest facing slopes to peat bogs and swamps. It occurs commonly in rather extreme moisture conditions, both very wet and quite dry. While many types of tree prefer a south or north facing aspect, the red maple does not appear to have a preference. Its ideal conditions are in moderately well-drained, moist sites at low or intermediate elevations. However, it is nonetheless common in mountainous areas on relatively dry ridges, as well as on both the south and west sides of upper slopes. Furthermore, it is common in swampy areas, along the banks of slow moving streams, as well as on poorly drained flats and depressions. In northern Michigan and New England, the tree is found on the tops of ridges, sandy or rocky upland and otherwise dry soils, as well as in nearly pure stands on moist soils and the edges of swamps. In the far south of its range, it is almost exclusively associated with swamps.

Interestingly, it is thought that the pre-European forest of eastern North America contained far fewer red maples than at present. Most diversity surveys conducted in eastern forests prior to their large scale exploitation showed the red maple representing under 5% of all tree species and it was furthermore mostly confined to poorly drained areas. The density of the tree in many of these areas has increased 6 to 7 fold and this trend seems to be continuing. A series of disturbances to the oak and pine forests since European arrival, such as the suppression of forest fires and global warming, are most likely responsible for this phenomenon. Concern has been expressed, as the ongoing spread of the red maple is changing the nature of eastern forests by reducing the number of oaks and pines that would otherwise dominate.

Ecology

Red maple seldom lives longer than 150 years, making it short to medium lived. It reaches maturity in 70 to 80 years. Its ability to thrive in a large number of habitats is largely due to its ability to produce roots to suit its site from a young age. In wet locations, red maple seedlings produce short taproots with long and developed lateral roots, while on dry sites, they develop long taproots with significantly shorter laterals. The roots are primarily horizontal, however, forming in the upper 25 cm (9.8 in) of the ground. Mature trees have woody roots up to 25 m (82 ft) long. They are very tolerant of flooding, with one study showing that 60 days of flooding caused no leaf damage. At the same time, they are tolerant of drought due to their ability to stop growing under dry conditions by then producing a second growth flush when conditions later improve, even if growth has stopped for 2 weeks. Red maple is one of the most drought-tolerant species of maple in the Carolinas.

A. rubrum is one of the first plants to flower in spring. A crop of seeds is generally produced every year with a bumper crop often occurring every second year. A single tree between 5 and 20 cm (2.0 and 7.9 in) in diametre can produce between 12,000 and 91,000 seeds in a season. A tree 30 cm (0.98 ft) in diametre was shown to produce nearly a million seeds. Red maple produces one of the smallest seeds of any of the maples. Fertilization has also been shown to significantly increase the seed yield for up to two years after application. The seeds are epigeal and tend to germinate in early Summer soon after they are released, assuming a small amount of light, moisture, and sufficient temperatures are present. If the seeds are densely shaded, then germination commonly does not occur until the next Spring. Most seedlings do not survive in closed forest canopy situations. However, one- to four-year-old seedlings are common under dense canopy and though they eventually die if no light reaches them, they serve as a reservoir, waiting to fill any open area of the canopy above.

Red maple is able to increase its numbers significantly when associate trees are damaged by disease, cutting, or fire. One study found that 6 years after clearcutting a 3.4 hectares (8.4 acres) Oak-Hickory forest containing no red maples, the plot contained more than 2,200 red maple seedlings per hectare (900 per acre) taller than 1.4 m (4.6 ft). One of its associates, the black cherry (*Prunus serotina*), contains benzoic acid, which has been shown to be a potential allelopathic inhibitor of red maple growth. Red maple is one of the first species to start stem elongation. In one study, stem elongation was one-half completed in 1 week, after which growth slowed and was 90% completed within only 54 days. In good light and moisture conditions, the seedlings can grow 30 cm (0.98 ft) in their first year and up to 60 cm (2.0 ft) each year for the next few years making it a fast grower.

The red maple is a used as a food source by several forms of wildlife. Elk and white-tailed deer in particular use the current season's growth of red maple as an important source of winter food. Several Lepidoptera (butterflies and moths) utilize the leaves as food; see List of Lepidoptera that feed on maples.

Due to *A. rubrum*'s very wide range, there is significant variation in hardiness, size, form, time of flushing, onset of dormancy, and other traits. Generally speaking, individuals from the north flush the earliest, have the most reddish Fall colour, set their buds the earliest and take the least winter injury. Seedlings are tallest in the north-central and east-central part of the range. The fruits also vary geographically with northern individuals in areas with brief frost free periods producing fruits that are shorter and heavier than their southern counterparts. As a result of the variation there is much genetic potential for breeding programmes with a goal of producing red maples for cultivation. This is especially useful for making urban cultivars that require resistance from verticillium wilt, air pollution, and drought.

Red maple frequently hybridizes with Silver Maple; the hybrid, known as Freeman's Maple *Acer x freemanii*, is intermediate between the parents.

Toxicity

The leaves of red maple, especially when dead or wilted, are extremely toxic to horses. The toxin is unknown, but believed to be an oxidant because it damages red blood cells, causing acute hemolysis that inhibits the transport of oxygen.

The ingestion of 700 grams (1.5 pounds) of leaves is considered toxic and 1.4 kilograms (3 pounds) is lethal. Symptoms occur within one or two days after ingestion and can include depression, lethargy, increased rate and depth of breathing, increased heart rate, jaundice, dark brown urine, coma, and death. Treatment is limited and can include the use of methylene blue or mineral oil and activated carbon in order to stop further absorption of the toxin into the stomach. About 50% to 75% of affected horses die or are euthanized as a result.

Cultivation

Red maple is widely grown as an ornamental tree in parks and large gardens, except where soils are too alkaline or salty. In parts of the Pacific Northwest, it is one of the most common introduced trees. Its popularity in cultivation stems from its vigorous habit, its attractive and early red flowers, and most importantly, its flaming red fall foliage. The tree was introduced into the United Kingdom in 1656 and shortly thereafter entered cultivation.

There it is frequently found in many parks and gardens, as well as occasionally in churchyards.

Red maple is a good choice of a tree for urban areas when there is ample room for its root system. The red maple is excellent at withstanding harsh urban conditions, including tolerance of both dry and wet soils, and a higher tolerance of pollution than sugar maple. Like several other maples, its low root system can be invasive and it makes a poor choice for plantings near paving. It attracts squirrels, who eat its buds in the early spring, although squirrels prefer the larger buds of the silver maple.

Cultivars

Numerous cultivars have been selected, often for intensity of fall colour, with 'October Glory' and 'Red Sunset' among the most popular. Towards its southern limit, 'Fireburst', 'Florida Flame', and 'Gulf Ember' are preferred. Many cultivars of the Freeman maple are also grown widely. Below is a partial list of cultivars:

- *'Armstrong'* - Columnar to fastigate in shape with silvery bark and modest orange to red fall foliage
- *'Autumn Blaze'* - Rounded oval form with leaves that resemble the silver maple. The fall colour is orange red and persists longer than usual
- *'Autumn Flame'* - A fast grower with exceptional bright red fall colour developing early. The leaves are also smaller than the species.
- *'Autumn Radiance'* - Dense oval crown with an orange-red fall colour
- *'Autumn Spire'* - Broad columnar crown; red fall colour; very hardy
- *'Bowhall'* - Conical to upright in form with a yellow-red fall colour
- *'Burgundy Bell'* - Compact rounded uniform shape with long lasting, burgundy fall leaves
- *'Columnare'* - An old cultivar growing to 20 metres (66 feet) with a narrow columnar to pyramidal form with dark green leaves turning orange and deep red in fall
- *'Gerling'* - A compact, slow growing selection, this individual only reaches 10 metres (33 feet) and has orange-red fall foliage
- *'Northwood'* - Branches are at a 45 degree angle to the trunk, forming a rounded oval crown. Though the foliage is deep green in summer, its orange-red fall colour is not as impressive as other cultivars.
- *'October Brilliance'* - This selection is slow to leaf in spring, but has a tight crown and deep red fall colour
- *'October Glory'* - Has a rounded oval crown with late developing intense red fall foliage. Along with 'Red Sunset', it is the most popular selection due to the dependable fall colour and vigorous growth. This cultivar has gained the Royal Horticultural Society's Award of Garden Merit.
- *'Red Sunset'* - The other very popular choice, this selection does well in heat due to its drought tolerance and has an upright

habit. It has very attractive orange-red fall colour and is also a rapid and vigorous grower.

- *'Scarlet Sentinel'* - A columnar to oval selection with 5-lobed leaves resembling the silver maple. The fall colour is yellow-orange to orange-red and the tree is a fast grower.
- *'Schlesingeri'* - A tree with a broad crown and early, long lasting fall colour that a deep red to reddish purple. Growth is also quite rapid.
- *'Shade King'* - This fast growing cultivar has an upright-oval form with deep green summer leaves that turn red to orange in fall.
- *'V.J. Drake'* - This selection is notable because the edges of the leaves first turn a deep red before the colour progresses into the centre.

Other Uses

In the lumber industry *Acer rubrum* is considered a soft maple. The wood is close grained and as such it is similar to that of *A. saccharum*, but its texture is softer, less dense, and has a poorer figure and machining qualities. High grades of wood from the red maple can nonetheless be substituted for hard maple, particularly when it comes to making furniture. As a soft maple, the wood tends to shrink more during the drying process than with the hard maples.

Red maple is also used for the production of maple syrup, though the hard maples *Acer saccharum* (sugar maple) and *Acer nigrum* (black maple) are more commonly utilized. One study compared the sap and syrup from the sugar maple with those of the red maple, as well as those of the *Acer saccharinum* (silver maple), *Acer negundo* (boxelder), and *Acer platanoides* (Norway maple), and all were found to be equal in sweetness, flavor, and quality. However, the buds of red maple and other soft maples emerge much earlier in the spring than the sugar maple, and after sprouting chemical makeup of the sap changes, imparting an undesirable flavor to the syrup. This being the case, red maple can only be tapped for syrup before the buds emerge, making the season very short.

Red maple is a medium quality firewood, possessing lower heat energy, nominally 5.4 MJ/m^3 (18.7 million BTU (mbtu) per cord), than other hardwoods such as Ash: 7 MJ/m^3 (24 mbtu/cord), Oak: 7 MJ/m^3 (24 mbtu/cord), or Birch: 5.8 MJ/m^3 (20 mbtu/cord).

Red Maple make vibrant and colourful Bonsai, and have year around attractive features for display.

Aesculus Hippocastanum

Aesculus hippocastanum is a large deciduous tree, commonly known as horse-chestnut or conker tree.

Distribution

Aesculus hippocastanum is native to a small area in the Pindus Mountains mixed forests and Balkan mixed forests of South East Europe. It is widely cultivated in streets and parks throughout the temperate world.

Growth

A. hippocastanum grows to 36 metres (118 ft) tall, with a domed crown of stout branches; on old trees the outer branches often pendulous with curled-up tips. The leaves are opposite and palmately compound, with 5–7 leaflets; each leaflet is 13–30 cm long, making the whole leaf up to 60 cm across, with a 7–20 cm petiole. The leaf scars left on twigs after the leaves have fallen have a distinctive horseshoe shape, complete with seven “nails”. The flowers are usually white with a small red spot; they are produced in spring in erect panicles 10–30 cm tall with about 20–50 flowers on each panicle. Usually only 1–5 fruit develop on each panicle; the shell is a green, spiky capsule containing one (rarely two or three) nut-like seeds called conkers or horse-chestnuts. Each conker is 2–4 cm diametre, glossy nut-brown with a whitish scar at the base.

Etymology

The common name “horse-chestnut” (often unhyphenated) is reported as having originated from the erroneous belief that the tree was a kind of chestnut (though in fact only distantly related), together with the observation that eating the fruit cured horses of chest complaints despite this plant being poisonous to horses.

Uses

Cultivation for its spectacular spring flowers is successful in a wide range of temperate climatic conditions provided summers are not too hot, with trees being grown as far north as Edmonton, Alberta, Canada, the Faroe Islands, and Harstad, Norway. This tree and the red flowering cultivar *A. hippocastanum* ‘Baumannii’ have gained the Royal Horticultural Society’s Award of Garden Merit.

In Britain and Ireland, the nuts are used for the popular children’s game conkers. During the two world wars, horse-chestnuts were used as a source of starch which in turn could be fermented via the *Clostridium acetobutylicum* method devised by Chaim Weizmann to produce acetone. This acetone was then used as a solvent which aided in the process of ballistite extrusion into cordite, which was then used in military armaments.

The nuts, especially those that are young and fresh, are slightly poisonous, containing alkaloid saponins and glucosides. Although not dangerous to touch, they cause sickness when eaten; consumed by horses, they can cause tremors and lack of coordination. Some mammals, notably deer, are able to break down the toxins and eat them safely.

In the past, horse-chestnut seeds were used in France and Switzerland for whitening hemp, flax, silk and wool. They contain a soapy juice, fit for washing of linens and stuffs, for milling of caps and stockings, etc., and for fulling of cloth. For this, 20 horse-chestnut seeds were sufficient for six litres of water. They were peeled, then rasped or dried, and ground in a malt or other mill.

The water must be soft, either rain or river water; hard well water will not work. The nuts are then steeped in cold water, which soon becomes frothy, as with soap, and then turns milky white. The liquid must be stirred well at first, and then, after standing to settle, strained or poured off clear. Linen washed in this liquid, and afterwards rinsed in clear running water, takes on an agreeable light sky-blue colour. It takes spots out of both linen and wool, and never damages or injures the cloth.

In Bavaria the chestnut is the typical tree for a beer garden. Originally they were planted for their deep shade which meant that beer cellar owners could cut ice from local rivers and lakes in winter to cool the Märzen Lager beer well into summer. Nowadays guests enjoy the shade to keep their heads cool.

Horse-chestnuts have been threatened by the leaf-mining moth *Cameraria ohridella*, whose larvae feed on horse chestnut leaves. The moth was described from Macedonia where the species was discovered in 1984 but took 18 years to reach Britain. The flower is the symbol of the city of Kiev, capital of Ukraine. Although the horse-chestnut is sometimes known as the buckeye, this name is generally reserved for the New World members of the *Aesculus* genus.

Medicinal Uses

The seed extract standardized to around 20 percent aescin (escin) is used for its venotonic effect, vascular protection, anti-inflammatory and free radical scavenging properties. Primary indication is chronic venous insufficiency. A recent Cochrane Review found the evidence suggests that Horse Chestnut Seed Extract is an efficacious and safe short-term treatment for chronic venous insufficiency.

Aescin reduces fluid leaks to surrounding tissue by reducing both the number and size of membrane pores in the veins.

Commonly used dose: Oral use, 250-300 mg extract standardized to around 20 percent aescin 1-3 times a day. Topical as a cream with 2% aescin.

Based on reports of worsening kidney function in people with kidney disease who received intravenous aescin, it is contraindicated in kidney and liver disease. Consistently very large oral doses should be avoided.

Raw Horse Chestnut seed, leaf, bark and flower are toxic due to the presence of esculin and should not be ingested. Horse chestnut seed is classified by the FDA as an unsafe herb. The glycoside and saponin constituents are considered toxic.

Aesculus hippocastanum is used in Bach flower remedies. When the buds are used it is referred to as "chestnut bud" and when the flowers are used it is referred to as "white chestnut".

Quercetin 3,4'-diglucoside, a flavonol glycoside can also be found in horse chestnut seeds.

Anne Frank Tree

A famous specimen of the horse-chestnut was the Anne Frank Tree in the centre of Amsterdam, which she mentioned in her diary and which survived until August 2010, when a heavy wind blew it over.

Bonsai

The horse-chestnut is a favourite subject for bonsai.

Diseases

- Bleeding Canker. Half of all horse-chestnuts in Great Britain are now showing symptoms to some degree of this potentially lethal bacterial infection.
- Guignardia leaf blotch, caused by the fungus *Guignardia aesculi*
- Wood rotting fungi, e.g. such as Armillaria and Ganoderma
- Horse-chestnut scale, caused by the insect *Pulvinaria regalis*
- Horse-chestnut leaf miner, *Cameraria ohridella*, a leaf mining moth. also affecting large numbers of UK trees.
- Phytophthora bleeding canker, a fungal infection.

Ageratina

Ageratina (snakeroot) is a genus of about 250–290 perennials and rounded shrubs in the family Asteraceae.

These plants grow mainly in the warmer regions of the Americas and West Indies. Over 150 species are native to Mexico. Some flourish in the cooler areas of the eastern United States. Two Mexican species have become a pest in parts of Australia and Taiwan. *Ageratina* used to belong to the genus *Eupatorium*, but it has been reclassified.

Description

The inflorescence consists of multiple fluffy, red or pinkish-white capitula in clusters. These lack the typical ray flowers of the composites.

They have multiple, much-branched woody stems. The petioles are rather long. The leaves are triangular, serrate and opposite with a foul-smelling, musky scent.

Species

- *Ageratina adenophora*
- *Ageratina altissima*
- *Ageratina amblyolepis*
- *Ageratina areolaris*
- *Ageratina aromatica*
- *Ageratina atrocordata*
- *Ageratina beamanii*
- *Ageratina bellidifolia*
- *Ageratina blepharilepis*
- *Ageratina brandegeana*
- *Ageratina brevipes*
- *Ageratina calaminthifolia*
- *Ageratina calophylla*
- *Ageratina campylocladia*
- *Ageratina capillipes*
- *Ageratina cardiophylla*
- *Ageratina cerifera*
- *Ageratina chazaroana*
- *Ageratina chiapensis*
- *Ageratina choricephala*

- *Ageratina collodes*
- *Ageratina conspicua*
- *Ageratina crassiramea*
- *Ageratina cremastra*
- *Ageratina cronquistii*
- *Ageratina cuencana*
- *Ageratina cylindrica*
- *Ageratina deltoidea*
- *Ageratina dendroides*
- *Ageratina dolichobasis*
- *Ageratina enixa*
- *Ageratina espinosarum*
- *Ageratina etlensis*
- *Ageratina flourensifolia*
- *Ageratina geminata*
- *Ageratina gentryana*
- *Ageratina glabrata*
- *Ageratina glauca*
- *Ageratina glischra*
- *Ageratina gonzalezorum*
- *Ageratina grashoffii*
- *Ageratina gypsophila*
- *Ageratina halbertiana*
- *Ageratina havanensis*
- *Ageratina hasegawana*
- *Ageratina helenae*
- *Ageratina henzium*
- *Ageratina herbacea*
- *Ageratina heterophylla*
- *Ageratina huahuapana*
- *Ageratina hyssopina*
- *Ageratina irrasa*
- *Ageratina isolepis*
- *Ageratina jaliscensis*

- *Ageratina jalpana*
- *Ageratina jolotepecana*
- *Ageratina josepaneroi*
- *Ageratina jucunda*
- *Ageratina kochiana*
- *Ageratina lasia*
- *Ageratina lasioneura*
- *Ageratina leiocarpa*
- *Ageratina lemmonii*
- *Ageratina leptodictyon*
- *Ageratina liebmannii*
- *Ageratina ligustrina*
- *Ageratina luciae-brauniae*
- *Ageratina lucida*
- *Ageratina macbridei*
- *Ageratina macdonaldii*
- *Ageratina macvaughii*
- *Ageratina mairetiana*
- *Ageratina malacolepis*
- *Ageratina manantlana*
- *Ageratina megacephala*
- *Ageratina miahuatlana*
- *Ageratina moorei*
- *Ageratina muelleri*
- *Ageratina neohintonorium*
- *Ageratina nesomii*
- *Ageratina oaxacana*
- *Ageratina occidentalis*
- *Ageratina oligocephala*
- *Ageratina oreithales*
- *Ageratina ovilla*
- *Ageratina ozolotepecana*
- *Ageratina palmeri*
- *Ageratina parayana*

- *Ageratina paupercula*
- *Ageratina pazcuarensis*
- *Ageratina pendula*
- *Ageratina pelotropha*
- *Ageratina petiolaris*
- *Ageratina photina*
- *Ageratina pichinchensis*
- *Ageratina potosina*
- *Ageratina pringlei*
- *Ageratina prunellifolia*
- *Ageratina pseudochilca*
- *Ageratina queretaroana*
- *Ageratina ramireziorum*
- *Ageratina ramonensis*
- *Ageratina resiniflua*
- *Ageratina rhomboidea*
- *Ageratina rhypodes*
- *Ageratina riparia*
- *Ageratina robinsoniana*
- *Ageratina rothrockii*
- *Ageratina rubicaulis*
- *Ageratina salicifolia*
- *Ageratina saltillensis*
- *Ageratina sandersii*
- *Ageratina scordonioides*
- *Ageratina shastensis*
- *Ageratina sodiroi*
- *Ageratina sousae*
- *Ageratina stricta*
- *Ageratina sundbergii*
- *Ageratina thyrsiflora*
- *Ageratina tomentella*
- *Ageratina triangulata*
- *Ageratina triniona*

- *Ageratina venulosa*
- *Ageratina vernalis*
- *Ageratina viburnoides*
- *Ageratina viejoana*
- *Ageratina villarrealii*
- *Ageratina viscosissima*
- *Ageratina warnockii*
- *Ageratina wrightii*
- *Ageratina yecorana*
- *Ageratina zunilana*

Toxicity

For more details on this topic, see Milk sickness. Milk from cows that have eaten snakeroot can cause illness if ingested because the milk becomes toxic. Symptoms of milk sickness include vomiting.

Medicinal Use

Ageratina pichinchensis is a traditional Mexican treatment for superficial fungal infections of the skin. These plant extracts contain encecalin which has activity to inhibit and kill the fungus. Studies have compared its effectiveness in treating toenail fungus with ciclopirox.

Aleurites

Aleurites is a small arborescent genus of flowering plants in the tropical and subtropical regions of Asia, the Pacific and South America, belonging to the spurge family Euphorbiaceae. These monoecious, evergreen trees are perennials or semi-perennials. These are large trees, 15–40 metres (49–130 ft) tall, with spreading drooping and rising branches.

The leaves are alternate, lobate, ovate to ovate-lanceolate with minute stipules. They are pubescent on both sides when young, but in a later stage they become glabrous.

The inflorescence consists of terminal plumes of small, creamy white bell-shaped fragrant flowers, branching from the base. The flowers are usually bisexual, with a solitary pistillate flower at the end of each major axis. The lateral cymes are staminate. There are five or six imbricate petals. The staminate flowers are mostly longer and thinner than the pistillate flowers, with 17-32 glabrous stamens in four whorls. The pistillate flowers have a superior ovary.

The fruits are rather large drupes with a fleshy exocarp and a thin, woody endocarp. They vary in shape, according to the numbers of developed locules. They contain oleiferous, poisonous seeds.

The oil has been used as a paraffin, lubricant and as a constituent of varnish, paint and soap. Once poisonous substances are removed, it can be used as a cooking oil. Some deciduous Chinese species are now classified under a separate genus *Vernicia*.

The name *Aleurites* is derived from a Greek word meaning "wheaten flour", because of the appearance of the lower surface of the leaf.

Species

The most widespread species is the Candlenut (*Aleurites moluccana*), occurring from tropical Asia, the Pacific, from India to China and Polynesia, Australia and New Zealand. Some botanists only recognize two species, *Aleurites moluccana* and *Aleurites rocking hamensis*.

- *Aleurites angustifolia* (New Caledonia)
- *Aleurites erratica*
- *Aleurites integrifolia* (New Caledonia)
- *Aleurites javanica* (Java)
- *Aleurites moluccana* (L.) Willd. – Indian walnut, candlenut tree, country walnut, aburagiri, ama
  * *Aleurites moluccana* var. *floccosa* (New Guinea)
  * *Aleurites moluccana* var. *katoi* O.Deg., I.Deg. & B.C.Stone (Hawaii)
  * *Aleurites moluccana* var. *moluccana*
  * *Aleurites moluccana* var. *rockinghamensis* (New Guinea, Australia)
- *Aleurites pentaphylla* (East Asia)
- *Aleurites rockinghamensis* (Baill.) P.I.Forst.
- *Aleurites remyi* (Hawaii)
- *Aleurites saponaria* (Philippines)
- *Aleurites triloba* (Pacific Islands)

Formerly Placed Here

- *Reutealis trisperma* (Blanco) Airy Shaw (as *A. trispermus* Blanco)

- *Vernicia cordata* (Thunb.) Airy Shaw (as *A. cordatus* (Thunb.) Müll.Arg.)
- *Vernicia fordii* (Hemsl.) Airy Shaw (as *A. fordii* Hemsl.)
- *Vernicia montana* Lour. (as *A. montanus* (Lour.) E.H.Wilson)

Amsinckia Menziesii

Amsinckia menziesii is a species of plant in the Boraginaceae family, the Borage or Forget-me-not family.

Varieties

The plant has two varieties:

Amsinckia menziesii var. *intermedia* - Common Fiddleneck, Intermediate Fiddleneck

Amsinckia menziesii var. *menziesii* - Menzies' Fiddleneck

Var. Intermedia

Amsinckia menziesii var. *intermedia* (Common Fiddleneck, or Intermediate Fiddleneck) is one of the common fiddlenecks of western North America, distributed from Alaska and Canada through the Western United States to Mexico.

Like other members of the genus, it has a terminal flowering whorl somewhat shaped like the head of a violin or fiddle, hence the name fiddleneck. The flowers are yellow-orange, orange, or dark yellow.

In Australia, the species has become a common weed of cultivated areas in New South Wales, Victoria and Queensland.

Its seeds are the favourite food of Lawrence's Goldfinch during that Californian bird's nesting season of spring and early summer.

Chapter 3

Medical Advances in Poisonous Plants

Erythrophleum Chlorostachys

Erythrophleum chlorostachys is a species of leguminous tree endemic to northern Australia, from northeastern Queensland to the Kimberley region of Western Australia. Commonly known as Cooktown Ironwood, the species is found in wide range of environments from arid savanna to tropical rainforest.

The species is a valuable source of timber, which is exceptionally hard and dense as well as being highly termite resistant.

Scientific Classification	
Kingdom:	Plantae
(unranked):	Angiosperms
(unranked):	Eudicots
(unranked):	
Order:	Fabales
Family:	Fabaceae
Genus:	*Erythrophleum*
Species:	*E. chlorostachys*

Ironwood is semi-deciduous, dropping much of its foliage in response to the prolonged winter dry periods which are the norm within its native range. The foliage of the tree contains toxic levels of alkaloids and has been responsible for numerous deaths of both cattle and horses.

Virtually all culturally modified trees in Eucalyptus tetradonta woodland on Cape York Peninsula are Cooktown ironwoods. Most of these are 'sugar bag scars' where Aboriginal people have cut through the cambium into the heartwood of the tree to remove honey from native bees. Scars have been made using both stone axes (in pre-contact times) and steel axes (post-contact).

These have particular significance to Aboriginal people as the tangible representation of past cultural practices. The large number of hollows found in Cooktown ironwoods at Kakadu National Park are also likely to be culturally modified trees.

Euphorbia

Euphorbia is a genus of flowering plants belonging to the family Euphorbiaceae. Consisting of 2008 species, *Euphorbia* is one of the largest and most diverse genera in the plant kingdom, along with *Rumex* and *Senecio*.

Members of the family and genus are commonly referred to as spurges. *Euphorbia antiquorum* is the type species for the genus *Euphorbia*; it was described by Carl Linnaeus in 1753 in *Species Plantarum*.

The family is primarily found in the tropical and subtropical regions of Africa and the Americas, but also in temperate zones worldwide. Succulent species originate mostly from Africa, the Americas and Madagascar. There exists a wide range of insular species: on the Hawaiian Islands, where spurges are collectively known as "akoko", and on the Canary Islands as "tabaibas".

The common name "spurge" derives from the Middle English/Old French *espurge* ("to purge"), due to the use of the plant's sap as a purgative.

The botanical name *Euphorbia* derives from Euphorbus, the Greek physician of king Juba II of Numidia (52–50 BC – 23 AD), who married the daughter of Anthony and Cleopatra. He wrote that one of the cactus-like Euphorbias was a powerful laxative. In 12 B.C., Juba named this plant after his physician Euphorbus in response to Augustus Caesar dedicating a statue to Antonius Musa, his own personal physician. Botanist and taxonomist Carl Linnaeus assigned the name *Euphorbia* to the entire genus in the physician's honour.

Description

The plants are annual or perennial herbs, woody shrubs or trees with a caustic, poisonous milky sap (latex). The roots are fine or thick and fleshy or tuberous. Many species are more or less succulent, thorny or unarmed.

The main stem and mostly also the side arms of the succulent species are thick and fleshy, 15–91 cm (6–36 inches) tall. The deciduous leaves are opposite, alternate or in whorls. In succulent species the leaves are mostly small and short-lived. The stipules are mostly small, partly transformed into spines or glands, or missing.

Like all members of the family Euphorbiaceae, all spurges have unisexual flowers. In *Euphorbia* these are greatly reduced and grouped into pseudanthia called cyathia. The majority of species are monoecious (bearing male and female flowers on the same plant), although some are dioecious with male and female flowers occurring on different plants. It is not unusual for the central cyathia of a cyme to be purely male, and for lateral cyathia to carry both sexes.

Sometimes young plants or those growing under unfavourable conditions are male only, and only produce female flowers in the cyathia with maturity or as growing conditions improve.

The bracts are often leaf-like, sometimes brightly coloured and attractive, sometimes reduced to tiny scales. The fruits are three (rarely two) compartment capsules, sometimes fleshy but almost always ripening to a woody container that then splits open (explosively). The seeds are 4-angled, oval or spherical, and in some species have a caruncle.

Xerophytes and Succulents

In the genus *Euphorbia*, succulence in the species has often evolved divergently and to differing degrees. Sometimes it is difficult to decide, and it is a question of interpretation, whether or not a species is really succulent or "only" xerophytic.

In some cases, especially with geophytes, plants closely related to the succulents are normal herbs. About 850 species are succulent in the strictest sense. If one includes slightly succulent and xerophytic species, this figure rises to about 1000, representing about 45% of all *Euphorbia* species.

Irritance

The milky sap of spurges (called "latex") evolved as a deterrent to herbivores. It is white and colourless when dry, except in *E. abdelkuri*, where it is yellow. The pressurized sap seeps from the slightest wound and congeals after a few minutes in air.

The specific composition of diterpene or triterpene esters varies, and determines how caustic and irritating a particular species is. In contact with mucous membranes (eyes, nose, mouth), the latex can produce extremely painful inflammation.

Therefore, spurges should be handled with caution and kept away from children and pets. Latex on skin should be washed off immediately and thoroughly. Congealed latex is insoluble in water, but can be removed with an emulsifier like milk or soap. A physician should be consulted if inflammation occurs, as severe eye damage including permanent blindness may result from exposure to the sap. When large succulent spurges in a greenhouse are cut, vapours can cause irritation to the eyes and throat several metres away. Precautions, including sufficient ventilation, are required.

Uses

Several spurges are grown as garden plants, among them Poinsettia (*E. pulcherrima*) and the succulent *E. trigona*. *E. pekinensis* (Chinese: 'Yb; pinyin: *dàjĐ*) is used in traditional Chinese medicine, where it is regarded as one of the 50 fundamental herbs. Several *Euphorbia* species are used as food plants by the larvae of some Lepidoptera (butterflies and moths), like the Spurge Hawk-moths (*Hyles euphorbiae* and *Hyles tithymali*), as well as the Giant Leopard Moth.

Systematics and Taxonomy

According to recent studies of DNA sequence data, most of the smaller "satellite genera" around the huge genus *Euphorbia* nest deep within the latter. Consequently these taxa, namely the never generally accepted genus *Chamaesyce* as well as the smaller genera *Cubanthus*, *Elaeophorbia*, *Endadenium*, *Monadenium*, *Synadenium* and *Pedilanthus* were transferred to *Euphorbia*. The entire subtribe Euphorbiinae now consists solely of the genus *Euphorbia*.

Selected Species

See List of *Euphorbia* species for complete list.

- *Euphorbia albomarginata* – Rattlesnake Weed, White-margined Sandmat
- *Euphorbia amygdaloides* – Wood Spurge
- *Euphorbia antisyphilitica* – Candelilla
- *Euphorbia balsamifera* – Sweet tabaiba (Canary Islands)
- *Euphorbia bulbispina*

- *Euphorbia calyptrata*
- *Euphorbia canariensis* – Cardón (Canary Islands)
- *Euphorbia caput-medusae* - Medusa's Head (South Africa)
- *Euphorbia characias* - Mediterranean Spurge
- *Euphorbia cyparissias* – Cypress Spurge
- *Euphorbia decidua*
- *Euphorbia dendroides* - Tree spurge
- *Euphorbia echinus*
- *Euphorbia elastica* – (Mexican) Palo Amarillo
- *Euphorbia epithymoides* – Cushion Spurge
- *Euphorbia esula* – Leafy Spurge
- *Euphorbia falcata*
- *Euphorbia franckiana*
- *Euphorbia grantii* – African Milk Bush
- *Euphorbia granulata*
- *Euphorbia griffithii*
- *Euphorbia guyoniana*
- *Euphorbia helioscopia* – Sun Spurge
- *Euphorbia heterophylla* – Painted Euphorbia, Desert Poinsettia, (Mexican) Fireplant, Paint Leaf, Kaliko
- *Euphorbia hirta* - Used in Philippines as a traditional dengue remedy
- *Euphorbia horrida* - African milk barrel
- *Euphorbia hypericifolia* - Diamond Frost (Inneuphe)
- *Euphorbia ingens* - Candelabra Tree
- *Euphorbia labatii*
- *Euphorbia lactea* – Mottled Spurge, Frilled Fan, Elkhorn
- *Euphorbia lathyris* – Caper Spurge, Paper Spurge, Gopher Spurge, Gopher Plant, Mole Plant
- *Euphorbia leuconeura* – Madagascar jewel
- *Euphorbia maculata* – Spotted Spurge, Prostrate Spurge
- *Euphorbia marginata* – Snow on the Mountain
- *Euphorbia mamillaris*
- *Euphorbia maritae*
- *Euphorbia milii* – Crown-of-thorns, Christ Plant

- *Euphorbia myrsinites* – Myrtle Spurge, Creeping Spurge, donkey tail
- *Euphorbia neriifolia* – Indian Spurge Tree
- *Euphorbia obesa*
- *Euphorbia obtusifolia*
- *Euphorbia paralias* – Sea Spurge
- *Euphorbia pekinensis* - Peking Spurge
- *Euphorbia peplis* – Purple Spurge
- *Euphorbia peplus* – Petty Spurge
- *Euphorbia polychroma* - Bonfire
- *Euphorbia pulcherrima* – Poinsettia, Mexican Flame Leaf, Christmas Star, Winter Rose, Noche Buena, Lalupatae, Pascua, *Atatürk çiçeði* (Turkish)
- *Euphorbia purpurea*
- *Euphorbia resinifera* – Resin Spurge
- *Euphorbia rigida* – Gopher Spurge, Upright Myrtle Spurge
- *Euphorbia serrata* – Serrated spurge, Sawtooth spurge
- *Euphorbia terracina* – Geraldton Carnation Weed
- *Euphorbia tirucalli* – Indian Tree Spurge, Milk Bush, Pencil Tree, Firestick
- *Euphorbia tithymaloides* – Devil's Backbone, "Redbird cactus", *cimora misha* (Peru)
- *Euphorbia virosa*

Subgenera

The genus *Euphorbia* is one of the largest and most complex genera of flowering plants and several botanists have made unsuccessful attempts to subdivide the genus into numerous smaller genera. According to the recent phylogenetic studies, *Euphorbia* can be divided into 4 subgenera, each containing several not yet sufficiently studied sections and groups.

Of these, *Esula* is the most basal. *Chamaesyce* and *Euphorbia* are probably sister taxa but very closely related to *Rhizanthium*. Extensive xeromorph adaptations in all probability evolved several times; it is not known if the common ancestor of the cactus-like *Rhizanthium* and *Euphorbia* lineages was xeromorphic—in which case a more normal morphology would have re-evolved namely in *Chamaesyce*—or whether extensive xeromorphism is entirely polyphyletic even to the level of the subgenera.

Festuca Arundinacea

Festuca arundinacea (syn., *Lolium arundinaceum* Darbyshire; *Schedonorus phoenix* (Scop.) Holub) is a species of fescue, commonly known as a type of grass called Tall fescue. It is a cool-season perennial C_3 species of bunchgrass native to Europe. It is an important forage grass throughout Europe, and many cultivars have been used in agriculture. It is also an ornamental grass in gardens, and a phytoremediation plant.

The predominant cultivar found in British pastures is S170, an endophyte-free variety. In its native European environment, tall fescue is found in damp grasslands, river banks, and in coastal seashore locations. Its distribution is a factor of climatic, edaphic, or other environmental attributes.

History

Tall fescue was introduced into the United States in the late 19th century, but it did not establish itself as a widely used perennial forage until the 1940s. As in Europe, tall fescue has become an important, well-adapted cool season forage grass for agriculture in the US with many cultivars. In addition to forage, it has become an important grass for turf and soil conservation. T

all fescue is the most heat tolerant of the major cool season grasses. Tall fescue has a deep root system compared to other cool season grasses. This non-native grass is well adapted to the "transition zone" Mid Atlantic and Southeastern United States and now occupies over 35,000,000 acres (140,000 km^2).

Tall fescue has become an invasive species and noxious weed in native California grasslands and habitats, such as the California coastal prairie plant community.

The dominant cultivar grown in the United States is Kentucky 31. In 1931 E. N. Fergus, a professor of agronomy at the University of Kentucky, collected seed from a population on a hillside in Menifee County, Kentucky although formal cultivar release did not happen until 1943. Fergus heard about this "wonder grass" while judging a sorghum syrup competition in a nearby town. He wanted to see this grass because it was green, lush, and growing well on a sloped hillside during a drought. While visiting the site he was impressed and took seed samples with him.

With this seed he conducted variety trials, initiated seed increase nurseries, and lauded its performance. It was released as Kentucky 31 in 1943 and today it dominates grasslands in the humid southeastern US. In 1943, Fergus and others recognized this tall fescue cultivar as being vigorous, widely adaptable, able to withstand poor soil conditions, resistant to pests and drought. It is used primarily in pastures and low maintenance situations.

Breeders have created numerous cultivars that are darker green with desirable narrower blades than the light green coarse bladed K-31. Tall fescue is the grass on the South Lawn.

Description

Tall fescue is a long-lived perennial bunchgrass species. Photosynthesis occurs throughout the leaves, which form bunches and are thick and wide with prominent veins running parallel the entire length of the blade. The blades have a “toothed” edge which can be felt if fingers are run down the edge of the leaf blade. The underside of the leaf may be shiny. Emerging leaves are rolled in the bud with no prominent ligule.

Note that most grasses are folded not rolled, which make this a key idenification feature on tall fescue. The auricles are usually blunt but occasionally may be more clawlike. The culm is round in cross-section. Typically, this species of grass has a long growing season and ranges between 2 to 4 feet (1.2 m) tall in seedhead stage.

Tall fescue spreads through tillering and seed transmission — not by stolon - above ground runnerss or rhizome - below ground runnerss, which are common in many grass species. However, tall fescue may have numerous sterile shoots that extend the width of each bunch. There are approximately 227,000 seeds per pound.

Typically found across the mid-atlantic and southeast US, tall fescue performs best in soils with pH values between 5.5 to 7. Growth may occur year-round if conditions are adequate, but typically growth ceases when soil temperature falls below 40 °F (4 °C).

Endophyte Association

Tall fescue can be found growing in most soils of the southeast including marginal, acidic, and poorly drained soils and in areas of low fertility, and where stresses occur due to drought and overgrazing. These beneficial attributes are now known to be a result of a symbiotic association with the fungus *Neotyphodium coenophialum*.

This association between tall fescue and the fungal endophyte is a mutualistic symbiotic relationship (both symbionts derive benefits from it). The fungus remains completely intercellular, growing between the cells of the aboveground parts of its grass host. The fungus is asexual, and is transmitted to new generations of tall fescue only through seed, a mode known as vertical transmission. Thus in nature, the fungus does not live outside the plant. Viability of the fungus in seeds is limited; typically, after a year or two of seed storage the fungal endophyte mycelium has died, and seeds germinated will result in plants that are endophyte-free.

The tall fescue–endophyte symbiosis confers a competitive advantage to the plant. Endophyte-infected tall fescue compared to endophyte-free tall fescue deters herbivory by insects and mammals, bestows drought resistance, and disease resistance. In return for shelter, seed transmission, and nutrients the endophyte produces secondary metabolites. These metabolites, namely alkaloids, are responsible for increased plant fitness. Alkaloids in endophytic tall fescue include 1-aminopyrrolizidines (lolines), ergot alkaloids (clavines, lysergic acids, and derivative alkaloids), and the pyrrolopyrazine, peramine.

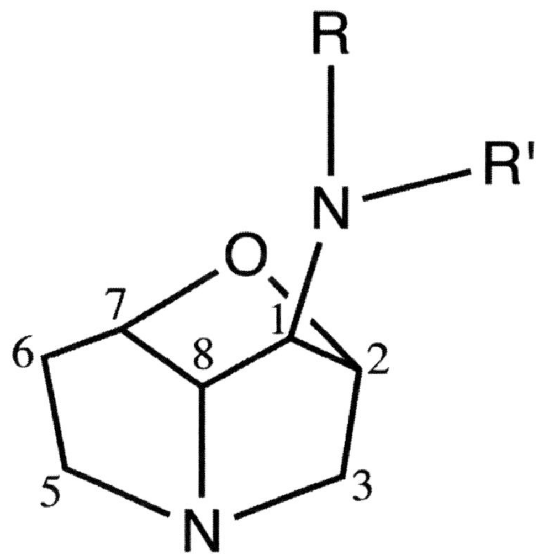

Figure: *Core structure of the 1-aminopyrrolizidines (loline alkaloids) produced in tall fescue infected by* Neotyphodium coenophialum*; R and R' denote variable substituents that can include methyl, formyl, and acetyl groups giving rise to different loline species.*

The lolines are the most abundant alkaloids, with concentrations 1000 higher than those of ergot alkaloids. Endophyte-free grasses do not produce lolines, and, as shown for the closely related endophyte commonly occurring in meadow fescue, *Neotyphodium uncinatum*, the endophyte can produce lolines in axenic laboratory culture.

However, although *N. coenophialum* possesses all the genes for loline biosynthesis, it does not produce lolines in culture. So in the tall fescue symbiosis, only the interaction of the host and endophyte produces the lolines. Lolines have been shown to deter insect herbivory, and may cause various other responses in higher organisms.

Despite their lower concentrations, ergot alkaloids appear to significantly affect animal growth. Ergots cause changes in normal homeostatic mechanisms in animals that result in toxicity manifested through reduced weight gains, elevated core temperatures, restricted blood flow, reduced milk production and reproductive problems. Peramine, like the ergot alkaloids, is found in much lower concentrations in the host compared with loline alkaloids. Its activity has been shown to be primarily insecticidal, and has not been linked to toxicity in mammals or other herbivores.

Endophyte Infected Tall Fescue Effects on Animals

Broodmares and Foals: Horses are especially prone to reproductive problems associated with tall fescue, often resulting in death of the foal, mare, or both. Horses which are pregnant may be strongly affected by alkaloids produced by the tall fescue symbiont. Broodmares that forage on infected fescue may have prolonged gestation, foaling difficulty, thickened placenta, or impaired lactation. In addition, the foals may be born weakened or dead. To moderate toxicosis, it is recommended that pregnant mares should be taken off infected tall fescue pasture for 60–90 days before foaling as late gestation problems are most common.

Cattle

Fescue toxicity in cattle appears as roughening of the coat in the summer and intolerance to heat. Cattle that graze on tall fescue are more likely to stay in the shade or wade in the water in hot weather. In the winter, a condition known as “fescue foot” might afflict cattle. This results from vasoconstriction of the blood vessels especially in the extremities, and causes a gangrenous condition. Untreated, the hoof might slough off. Additionally, cattle may experience decreased weight gains and poor milk production when heavily grazing infected tall fescue pasture. To deter toxicosis cattle should be given alternative feed to dilute their infected tall fescue intake.

Nutrient Pools Under Tall Fescue Pasture

Carbon cycling in terrestrial ecosystems is a major focus of research. Terrestrial carbon sequestration is the process of removing carbon dioxide from the atmosphere via photosynthesis and storing this carbon in either plant or soil carbon pools.

Increases in soil organic carbon help aggregate the soil, increase infiltration, reduce erosion, increase soil fertility, and act as long lived pools of soil carbon.

Many studies have suggested that long term endophyte-infected tall fescue plots increase soil carbon storage in the soil by limiting the microbial and macrofaunal activity to break down endophyte infected organic matter input and by increasing inputs of carbon via plant production. While the long term studies tend to show an increase in carbon storage, the short term studies do not.

However, short term studies have shown that the endophyte association results in higher above- and belowground plant biomass production compared to uninfected plants, as well as a decrease in certain microbial communities.

Site-specific characteristics, such as management and climate, need to be further understood to realize the ecological role and potential benefits of tall fescue and the endophyte association as it relates to carbon sequestration.

Novel Endophytes

New cultivars are being bred and tested every year. A major focus of research is producing endophyte-infected tall fescue cultivars that have no detrimental effects to livestock while keeping the endophytic effects of reduced insect herbivory, disease resistance, drought tolerance, and extended growing season.

Novel endophytes, also referred to as "friendly" endophytes, are symbiotic fungi that are associated with tall fescue, but do not produce target alkaloids in toxic concentrations. A widely used and tested novel endophyte is called MaxQ and is grown in the tall fescue grass host Georgia-Jesup. This cultivar of tall fescue-novel endophyte combination produces ergot alkaloids at near zero levels while maintaining the concentration of other alkaloids.

Tall Fescue Information System

More information can be found on the Tall Fescue Information System (TFIS). This resource provides up to the date, peer reviewed information relating to tall fescue.

Ambrosia Tomentosa

Ambrosia tomentosa (Skeletonleaf bur ragweed, Silverleaf povertyweed, Skeleton-leaf bursage) is a perennial species in the Aster family (Asteraceae). This plant grows up to 3 feet (91 cm) tall. The deeply lobed hairy leaves grow to 5 inches (12.7 cm) and have toothed margins. Flowers are small and yellow and produce spined 2-seeded burrs. In addition to seeds it can also reproduce via its widely spreading roots.

Skeletonleaf bur ragweed is native to the Central and Western United States. It is considered a noxious weed in several states.

Glechoma Hederacea

Glechoma hederacea (syn. *Nepeta glechoma* Benth., *Nepeta hederacea* (L.) Trevir.) is an aromatic, perennial, evergreen creeper of the mint family Lamiaceae. It is commonly known as Ground-ivy, gill-over-the-ground or Creeping Charlie. It has numerous medicinal uses, and is commonly used as a salad green in many countries. European settlers carried it around the world, and it has become a well established naturalized/introduced plant in a wide variety of localities.

Glechoma hederacea is native to Europe and southwestern Asia but has been introduced to North America and is now common in most regions other than the Rocky Mountains. Its common names include Alehoof, Creeping Charlie (or Charley), Catsfoot (from the size and shape of the leaf), Field Balm, Run-away-robin, Ground Ivy, Gill-over-the-ground, and Tunhoof. It is also sometimes known as *Creeping Jenny*, but that more commonly refers to *Lysimachia nummularia*.

It can be identified by its round to reniform (kidney or fan shaped), crenate (with round toothed edges) opposed leaves 2–3 cm diametre, on 3–6 cm long petioles attached to square stems which root at the nodes. It is a variable species, its size being influenced by environmental conditions, from 5 cm up to 50 cm tall.

Glechoma is sometimes confused with common mallow or *Malva neglecta*, which also has round, lobed leaves; but mallow leaves are attached to the stem at the back of a rounded leaf, where ground ivy has square stems and leaves which are attached in the centre of the leaf, more prominent rounded lobes on their edges, attach to the stems in an opposite arrangement, and have a hairy upper surface.

In addition, mallow and other creeping plants sometimes confused with ground ivy do not spread from nodes on stems. In addition, ground ivy emits a distinctive odor when damaged, being a member of the mint family.

The flowers of *Glechoma* are bilaterally symmetrical, funnel shaped, blue or bluish-violet to lavender, and grow in opposed clusters of 2 or 3 flowers in the leaf axils on the upper part of the stem or near the tip. It usually flowers in the spring.

Glechoma thrives in moist shaded areas, but also tolerates sun very well. It is a common plant in grasslands and wooded areas or wasteland. It also thrives in lawns and around buildings since it survives mowing.

It spreads by stolons or by seed. Part of the reason for its wide spread is this rhizomatous method of reproduction. It will form dense mats which can take over areas of lawn, and thus can be considered potentially invasive or aggressive weed.

Ecological Aspects

A number of wild bees fly upon this plant, including *Anthophora furcata*, *Anthidum manicatum*, *Anthophora plumipes*, *Anthophora quadrimaculata*, *Osmia aurulenta*, *Osmia caerulentes*, and *Osmia uncinata*. The plant is also galled by several insects, including *Rondaniola bursaria* (Lighthouse Gall), *Liposthenes glechomae* or *Liposthenes latreillei* (Kieffer, 1898) (a gall wasp).

Cultivation and Medicinal and Culinary Uses

Glechoma is quite attractive. It is grown as a potted plant and occasionally as a ground cover. Easily cultivated, it grows well in shaded places. A variegated variety is commercially available; in many areas this is the dominant form which has escaped cultivation and become established as an aggressive, adventitious ground cover.

While often thought of as a weed because of its propensity for spreading, *Glechoma* has culinary and medicinal uses which were the cause of its being imported to America by early European settlers. The fresh herb can be rinsed and steeped in hot water to create an herbal tea which is rich in vitamin C. It has a distinctive, mildly peppery flavor; it can be cooked as a pot herb, although it is most commonly eaten as a fresh salad green.

Glechoma was also widely used by the Saxons in brewing beer as flavoring, clarification, and preservative, before the introduction of hops for these purposes; thus the brewing-related names, Alehoof, Tunhoof, and Gill-over-the-ground.

Glechoma has been used in the cheese-making process as a substitute for animal rennet.

Traditional Medicine

Glechoma hederacea has been used in the traditional medicine of Europe going back thousands of years. Galen recommends the plant to treat inflammation of the eyes. John Gerard, an English herbalist, recommended the plant to treat tinnitus, as well as a "diuretic, astringent, tonic and gentle stimulant. Useful in kidney diseases and for indigestion."

It has also been used as a "lung herb". Other traditional uses include as an expectorant, astringent, and to treat bronchitis. The essential oil of the plant has been used for centuries as a general tonic for colds and coughs, and to relieve congestion of the mucous membranes.

Safety

Although it has been used as a salad green and in herbal medicines for thousands of years, the safety of *Glechoma hederacea* has not been established scientifically, and there is sufficient evidence to warrant caution with its use. *Glechoma hederacea* is toxic to cattle and horses. *Glechoma hederacea* is known to contain terpenoids; terpene-rich volatile oils are known to irritate the gastrointestinal tract and kidneys. The volatile oil also contains pulegone, a chemical also occurring in pennyroyal that is a known irritant, toxic to the liver, and also an abortifacient. The total yield of volatile oil in *Glechoma* is less than 1/30th the concentration that of pennyroyal.

Control

As is often the case when a plant has this many familiar names, *Glechoma* is familiar to a large number of people as a weed, a property it shares with many others of the mint family. It can be a problem in heavy, rich soils with good fertility, high moisture, and low boron content. It thrives particularly well in shady areas where grass does not grow well, although it can also be a problem in full sun.

Small infestations can be controlled through hand weeding; repeated weeding is required because the plant is stoloniferous and will continue to spread from its roots or bits of stem which reroot.

Glechoma is unusually sensitive to boron, and can be killed by applying borax (sodium tetraborate) in solution. The ratio is eight to ten ounces (225 - 275gm) of borax dissolved in four ounces (125ml) of warm water, diluted to 2.5 U.S. gallons (10 ltrs) of final solution, to be sprayed evenly over precisely 1,000 square feet (100 m^2) of lawn "no more, no less".

Note that despite being a "natural" treatment, boron is toxic to other plants and to animals at only slightly higher concentrations and, being an element, does not break down; therefore the long term effects of this technique on soil or groundwater, although not well documented, can be assumed to be unfavourable. More recent research discounts the efficacy of borax, primarily because finding the correct concentration for a given area is difficult and the potential for damaging desired plants.

Aside from mechanical removal or borax treatment, the other alternative for *Glechoma* infestation is use of commercial herbicides. There is some disagreement over the effectiveness of various herbicides, with dicamba (Trimec and Weed-B-Gon) and 2,4-D being described variously as both effective and ineffective by different sources. Some or all of the disagreement may be due to the existence of subpopulations which have differing susceptibilities to different compounds, as well as to differing rates of application.

To avoid generating herbicide resistance, the same product should not be used several years in succession; rather, various products should be used in rotation. Triclopyr has also been described as effective, and Clopyralid, MCPP, and quinclorac as ineffective. Fluroxypyr and Confront have also been described as effective, but sales of both are restricted to professionals. Two applications ten to fourteen days apart are necessary; also, the ability of the surviving plants to regenerate after 24 days can require a second treatment four or five weeks later, and even more followups.

In addition, the timing of application may play a role in the effectiveness of the herbicide, as well as the perception of effectiveness. For instance, fall is usually the best time for use of broadleaf herbicides; however a slow acting herbicide like triclopyr applied in the fall may not appear to have been effective until the next growing season.

Other techniques reported effective are to fertilize with greater than two pounds of nitrogen per thousand square feet annually, and use of the preemergence herbicide, isoxaben.

In extremely difficult cases, a short-lived full-spectrum herbicide such as Roundup is used to kill the entire lawn, and it is reseeded from start.

Grindelia

Grindelia is a genus of plants native to the Americas belonging to the family Asteraceae, (Compositae). *G. squarrosa*, a plant with bright yellow flowers indigenous to much of the United States, is commonly called curlycup gumweed. *G. robusta*, found in the western states, is a coastal scrub bush that is reputed to have several medicinal uses.

Hairy gumweed, *Grindelia cuneifolia*, occurs in brackish coastal marshes of western North America, such as in some portions of the San Francisco Bay perimetre.

Grindelia species are used as food plants by the larvae of some Lepidoptera species including *Schinia mortua*.

Selected species:

- *Grindelia buphthalmoides*
- *Grindelia camporum*
- *Grindelia chiloensis*
- *Grindelia ciliata* - goldenweed
- *Grindelia fraxino-pratensis* - Ash Meadows gumplant
- *Grindelia hirsutula*
- *Grindelia integrifolia* - Puget Sound gumweed
- *Grindelia nana* - Idaho gumplant
- *Grindelia squarrosa*
- *Grindelia stricta* - Oregon gumweed

Heliotropium

Heliotropium is a genus of flowering plants in the borage family, Boraginaceae. There are 250 to 300 species in this genus, which are commonly known as heliotropes.

Ecology and Human Use

Several heliotropes are popular garden plants, most notably Garden Heliotrope (*H. arborescens*). Some species are weeds and many are hepatotoxic if eaten in large quantities due to abundant pyrrolizidine alkaloids.

Danainae butterflies like to visit these plants, as pyrrolizidine alkaloids produce a kind of "perfume" to attract mates.

Caterpillars of the Grass Jewel (*Freyeria trochylus*), a gossamer-winged butterfly, feed on *H. strigosum*.

The sap of heliotrope flowers, namely of European Heliotrope (*H. europaeum*), was used as a food colouring in Middle Ages and Early Modern French cuisine.

One of the most famous ragtime piano melodies is *Heliotrope Bouquet*, composed in 1907 by Louis Chauvin (the first two strains) and Scott Joplin (the last two strains). Garden Heliotrope is grown in Southern Europe as an ingredient for perfume.

Selected Species

- *Heliotropium amplexicaule* Vahl – Clasping Heliotrope, Summer Heliotrope, Blue Heliotrope
- *Heliotropium anderssonii*
- *Heliotropium anomalum* Hook. & Arn. – Polynesian Heliotrope, Pacific Heliotrope (Pacific Islands)
    - * *Heliotropium anomalum* var. *argenteum* – *Hinahina kû kahakai* (Hawaii)
- *Heliotropium arborescens* – Garden Heliotrope, Common Heliotrope, "cherry pie"
- *Heliotropium argenteum*
- *Heliotropium asperrimum* R.Br.
- *Heliotropium balfourii*
- *Heliotropium bracteatum* R.Br.
- *Heliotropium conocarpum* F.Muell. ex Benth.
- *Heliotropium crispatum* F.Muell. ex Benth.
- *Heliotropium diversifolium* F.Muell. ex Benth.
- *Heliotropium chenopodiaceum* (A.DC.) Clos.
- *Heliotropium claussenii* DC.
- *Heliotropium convolvulaceum* – Sweet-scented Heliotrope, Showy Heliotrope
- *Heliotropium curassavicum* L. – Seaside Heliotrope, Salt Heliotrope, Monkey Tail, Quail Plant, "Chinese Parsley"; *Cola de Mico* (Spanish)
- *Heliotropium dentatum*
- *Heliotropium derafontense*
- *Heliotropium ellipticum*
- *Heliotropium epacrideum* F.Muell. ex Benth.
- *Heliotropium europaeum* L. – European Heliotrope, European Turnsole (Europe, Asia, and North Africa)
- *Heliotropium fasciculatum* R.Br.
- *Heliotropium flintii* F.Muell. ex A.S.Mitch.

- *Heliotropium foertherianum* Diane & Hilger – Tree Heliotrope, Velvet Soldierbush, Octopus Bush (South Asia, East Asia, Melanesia, western Polynesia, northern Australia)
- *Heliotropium foliatum* R.Br.
- *Heliotropium glabellum* R.Br.
- *Heliotropium heteranthum* (F.Muell.) Ewart & O.B.Davies
- *Heliotropium indicum* L. – Indian Turnsole
- *Heliotropium kuriense*
- *Heliotropium laceolatum* Loefg.
- *Heliotropiumleariifolium* Phil.
- *Heliotropium megalanthumn* I.M.Johnst.
- *Heliotropium nigricans*
- *Heliotropium paniculatum* R.Br.
- *Heliotropium pannifolium* – St. Helena Heliotrope (Saint Helena) (extinct, c.1820)
- *Heliotropium pauciflorum* R.Br.
- *Heliotropium paulayanum*
- *Heliotropium peruvianum*
- *Heliotropium pleiopterum* F.Muell.
- *Heliotropium procumbens*
- *Heliotropium prostratum* R.Br.
- *Heliotropium riebeckii*
- *Heliotropium shoabense*
- *Heliotropium sinuatum* (Miers) I.M.Johnst.
- *Heliotropium socotranum*
- *Heliotropium stenophyllum*
- *Heliotropium strigosum* Willd.
- *Heliotropium tenuifolium* R.Br.
- *Heliotropium ventricosum* R.Br.
- *Heliotropium wagneri*
- *Heliotropium* aff. *wagneri* (Samhah, Yemen)

Formerly included here

- *Chrozophora tinctoria* (as *H. tricoccum*)
- *Tournefortia gnaphalodes* (L.) R.Br. ex Roem. & Schult. (as *H. gnaphalodes* L.)

Homeria

Homeria, commonly known as Cape tulips, is a genus of the botanical family Iridaceae, which has 32 species of corm-bearing perennial herbs. In addition, these plants are used as ornamental plants. In the early 19th century, this genus was recognized as a single species of tulip, *Tulipa breyniana*. Currently, this genus is considered a synonym of *Moraea*.

St John's Wort

St John's wort is the plant species *Hypericum perforatum*, and is also known as Tipton's weed, rosin rose, goatweed, chase-devil, or Klamath weed.

With qualifiers, St John's wort is used to refer to any species of the genus *Hypericum*. Therefore, *H. perforatum* is sometimes called common St John's wort or perforate St John's wort to differentiate it. The species of *Hypericum* are classified in the Hypericaceae family, having previously been classified as Guttiferae or Clusiaceae. Approximately 370 species of the genus *Hypericum* exist worldwide with a native geographical distribution including temperate and subtropical regions of North America, Europe, Turkey, Russia, India, China and Brazil.

St John's wort is widely known as an herbal medicine for treating mild forms of depression.

Botanical Description

Hypericum perforatum is a yellow-flowering, stoloniferous or sarmentose, perennial herb indigenous to Europe, which has been introduced to many temperate areas of the world and grows wild in many meadows.

The common name comes from its traditional flowering and harvesting on St John's day, 24 June. The genus name *Hypericum* is derived from the Greek words *hyper* (above) and *eikon* (picture), in reference to the traditional use of the plant to ward off evil, by hanging plants over a religious icon in the house during St John's day. The species name *perforatum* refers to the presence of small oil glands in the leaves that look like windows, which can be seen when they are held against the light.

St John's wort is a perennial plant with extensive, creeping rhizomes. Its stems are erect, branched in the upper section, and can grow to 1 m high. It has opposing, stalkless, narrow, oblong leaves that are 12 mm long or slightly larger. The leaves are yellow-green in colour, with transparent dots throughout the tissue and occasionally with a few black dots on the lower surface. Leaves exhibit obvious translucent dots when held up to the light, giving them a 'perforated' appearance, hence the plant's Latin name.

Its flowers measure up to 2.5 cm across, have five petals, and are coloured bright yellow with conspicuous black dots. The flowers appear in broad cymes at the ends of the upper branches, between late spring and early to mid summer.

The sepals are pointed, with glandular dots in the tissue. There are many stamens, which are united at the base into three bundles. The pollen grains are ellipsoidal.

When flower buds (not the flowers themselves) or seed pods are crushed, a reddish/purple liquid is produced.

Ecology

St John's wort reproduces both vegetatively and sexually. It thrives in areas with either a winter- or summer-dominant rainfall pattern; however, distribution is restricted by temperatures too low for seed germination or seedling survival. Altitudes greater than 1500 m, rainfall less than 500 mm, and a daily mean January (in Southern hemisphere) temperature greater than 24 degrees C are considered limiting thresholds.

Depending on environmental and climatic conditions, and rosette age, St John's wort will alter growth form and habit to promote survival. Summer rains are particularly effective in allowing the plant to grow vegetatively, following defoliation by insects or grazing.

The seeds can persist for decades in the soil seed bank, germinating following disturbance.

Invasive Species

Although *Hypericum perforatum* is grown commercially in some regions of south east Europe, it is listed as a noxious weed in more than twenty countries and has introduced populations in South and North America, India, New Zealand, Australia, and South Africa.

In pastures, St John's wort acts as both a toxic and invasive weed. It replaces native plant communities and forage vegetation to the dominating extent of making productive land nonviable or becoming an invasive species in natural habitats and ecosystems.

Ingestion by livestock can cause photosensitization, central nervous system depression, spontaneous abortion, and can lead to death. Effective herbicides for control of *Hypericum* include 2,4-D, picloram, and glyphosate. In western North America three beetles *Chrysolina quadrigemina*, *Chrysolina hyperici* and *Agrilus hyperici* have been introduced as biocontrol agents.

Medicinal Uses

Depression Treatment Use

St John's wort is widely known as an herbal treatment for depression. In some countries, such as Germany, it is commonly prescribed for mild depression, especially in children and adolescents.

It is proposed that the mechanism of action of St. John's wort is due to the inhibition of reuptake of certain neurotransmitters.

A report from the Cochrane Review states:

The available evidence suggests that the *Hypericum* extracts tested in the included trials a) are superior to placebo in patients with major depression; b) are similarly effective as standard antidepressants; and c) have fewer side-effects than standard antidepressants.

There are two issues that complicate the interpretation of our findings:

1. While the influence of precision on study results in placebo-controlled trials is less pronounced in this updated version of our review compared to the previous version (Linde 2005a), results from more precise trials still show smaller effects over placebo than less precise trials.
2. Results from German-language countries are considerably more favourable for Hypericum than trials from other countries.

In one study, St John's wort was not found to be effective for patients suffering from dysthymia, a less severe and more chronic variety of depression. However, St John's wort has been shown to be effective in the treatment of mild to moderate depression in certain studies.

Standardized extracts are generally available over the counter, though in some countries (such as the Republic of Ireland) a prescription is required. Extracts are usually in tablet or capsule form, and also in teabags and tinctures.

Major Depressive Disorder

An analysis of twenty-nine clinical trials with more than five thousand patients was conducted by Cochrane Collaboration. The review concluded that extracts of St John's wort were superior to placebo in patients with major depression. St John's wort had similar efficacy to standard antidepressants. The rate of side-effects was half that of newer SSRI antidepressants and one-fifth that of older tricyclic antidepressants.

The National Centre for Complementary and Alternative Medicine (NCCAM) and other NIH-affiliated organizations hold that St John's wort has minimal or no effects beyond placebo in the treatment of major depression. This conclusion is based primarily on one trial of 340 volunteers, with negative outcome conducted by NCCAM. The authors of the study themselves, as well as several others, pointed out the low assay sensitivity of this study, and how only limited conclusions can be drawn from its results. The same study also indicated that sertraline (Zoloft) has no positive effects vs the same placebo.

A study published in 2005 found that St John's wort was significantly more effective than fluoxetine (Prozac).

Dosage

Most studies of St John's wort for treating depression used doses of 300 mg of an extract (standardized to 0.3 percent hypericin content) three times daily to achieve a therapeutic effect. Studies suggest to take from 300 mg to 1,800 mg daily. There have not been any reports of overdose.

Other Medical Uses

St John's wort is being studied for effectiveness in the treatment of certain somatoform disorders. Results from the initial studies are mixed and still inconclusive; some research has found no effectiveness, other research has found a slight lightening of symptoms. Further study is needed and is being performed.

A major constituent chemical, hyperforin, may be useful for treatment of alcoholism, although dosage, safety and efficacy have not been studied. Hyperforin has also displayed antibacterial properties against gram-positive bacteria, although dosage, safety and efficacy has not been studied. Herbal medicine has also employed lipophilic extracts from St John's wort as a topical remedy for wounds, abrasions, burns, and muscle pain.

The positive effects that have been observed are generally attributed to hyperforin due to its possible antibacterial and anti-inflammatory effects. For this reason hyperforin may be useful in the treatment of infected wounds and inflammatory skin diseases. In response to hyperforin's incorporation into a new bath oil, a study to assess potential skin irritation was conducted which found good skin tolerance of St John's wort.

A randomized controlled trial of St John's wort found no significant difference between it and placebo in the management of ADHD symptoms over eight weeks. However, the St John's wort extract used in the study, originally confirmed to contain 0.3% hypericin, was allowed to degrade to levels of 0.13% hypericin and 0.14% hyperforin. Given that the level of hyperforin was not ascertained at the beginning of the study, and levels of both hyperforin and hypericin were well below that used in other studies, little can be determined based on this study alone.

Hypericin and pseudohypericin have shown both antiviral and antibacterial activities. It is believed that these molecules bind non-specifically to viral and cellular membranes and can result in photo-oxidation of the pathogens to kill them.

A research team from the Universidad Complutense de Madrid (UCM) published a study entitled "*Hypericum perforatum*. Possible option against Parkinson's disease", which suggests that St John's wort has antioxidant active ingredients that could help reduce the neuronal degeneration caused by the disease.

Recent evidence suggests that daily treatment with St John's wort may improve the most common physical and behavioural symptoms associated with premenstrual syndrome.

St John's wort was found to be less effective than placebo, in a randomized, double-blind, placebo-controlled trial, for the treatment of irritable bowel syndrome.

St John's wort alleviated age-related long-term memory impairment in rats.

Chemical Constituents

Herb and flowers contain various polyphenols: flavonoids (epigallocatechin, rutin, hyperoside, isoquercetin, quercitrin, quercetin, I3, II8-biapigenin, amentoflavone, astilbin, miquelianin), phenolic acids (chlorogenic acid, 3-O-coumaroylquinic acid), and various naphtodianthrones: (hypericin, pseudohypericin, protohypericin, protopseudohypericin), phloroglucinols (hyperforin, adhyperforin). The naphthodianthrones hypericin and pseudohypericin along with the phloroglucinol derivative hyperforin are thought to be the active components. It also contains essential oils composed mainly of sesquiterpenes.

Pharmacology

The exact mechanism by which St John's wort functions is unclear and subject to conjecture.

The St John's wort mechanism is believed to involve inhibition of serotonin (5 HT) reuptake, much like the conventional selective serotonin reuptake inhibitor (SSRI) antidepressants. The major active antidepressive constituents in St John's wort are thought to be hyperforin and hypericin, although other biologically active constituents present, for example, flavonoids and tannins, may also be involved.

Some believe that hyperforin is the major constituent responsible for antidepressant activity, and it has been shown to inhibit the uptake of 5-HT, dopamine, and noradrenaline. Hyperforin also has affinity for GABA and glutamate transporters. On the other hand, a hyperforin-free extract of St John's wort (Ze 117 – Remotiv) may still have significant antidepressive effects.

Adverse Effects and Drug Interactions

St John's wort is generally well tolerated, with an adverse effect profile similar to placebo. The most common adverse effects reported are gastrointestinal symptoms, dizziness, confusion, tiredness and sedation.

It also decreases the levels of estrogens, such as estradiol, by speeding up its metabolism, and should not be taken by women on contraceptive pills as it upregulates the P450 system in the liver.

St John's wort may rarely cause photosensitivity. This can lead to visual sensitivity to light and to sunburns in situations that would not normally cause them. Related to this, recent studies concluded that the extract reacts with light, both visible and ultraviolet, to produce free radicals, molecules that can damage the cells of the body.

These can react with vital proteins in the eye that, if damaged, precipitate out, causing cataracts. This finding is contradicted by the results of another recent study, which found that St John's wort inhibits free radical production in both cell-free and human vascular tissue, revealing antioxidant properties of the compound.

St John's wort is associated with aggravating psychosis in people who have schizophrenia.

Pharmacokinetic Interactions

St John's wort has been shown to cause multiple drug interactions through induction of the cytochrome P450 enzyme CYP3A4, but also

CYP2C19. This results in the increased metabolism of those drugs, resulting in decreased concentration and clinical effect. The principal constituents thought to be responsible are hyperforin and amentoflavone.

St John's wort also has been shown to cause drug interactions through the induction of the P-glycoprotein (P-gp) efflux transporter. Increased P-gp expression results in decreased absorption and increased clearance of those drugs, which leads to lower clinical concentrations and efficacy.

Table: *Examples of drugs causing clinically significant interactions with St John's wort*

Class	***Drugs***
Antiretrovirals	Non-nucleoside reverse transcriptase inhibitors, protease inhibitors
Benzodiazepines	Alprazolam, midazolam
Hormonal contraception	Combined oral contraceptives
Immunosuppressants	Calcineurin inhibitors, cyclosporine, tacrolimus
Antiarrhythmics	Amiodarone, flecainide, mexiletine
Beta-blockers	Metoprolol, carvedilol
Calcium channel blockers	Verapamil, diltiazem, amlodipine
Statins (cholesterol-reducing medications)	Lovastatin, simvastatin, atorvastatin
Others	Digoxin, methadone, omeprazole, phenobarbital, theophylline, warfarin, levodopa, buprenorphine, irinotecan

Reference: Rossi, 2005; Micromedex

For a complete list, see CYP3A4 ligands and CYP2C9 ligands. For further updating on interactions and appropriate management, see Herbological.com – St John's Wort Interactions table (outdated since 2005).

Pharmacodynamic Interactions

In combination with other drugs that may elevate 5-HT (serotonin) levels in the central nervous system (CNS), St John's wort may contribute to serotonin syndrome, a potentially life-threatening adverse drug reaction.

Drugs that may contribute to serotonin syndrome with St John's wort

Class	Drugs
Antidepressants	MAOIs, TCAs, SSRIs, SNRIs, mirtazapine
Opioids	Tramadol, meperidine (pethidine), dextromethorphan, Levorphanol
CNS stimulants	Phentermine, diethylpropion, amphetamines, sibutramine, cocaine
5-HT$_1$ agonists	Triptans
Psychedelic drugs	Methylenedioxymethamphetamine (MDMA), lysergic acid diethylamide (LSD), psilocybin / psilocin, Mescaline and virtually every serotonergic psychedelic.
Others	Selegiline, tryptophan, buspirone, lithium, linezolid, 5-HTP

Reference: Rossi, 2005

Detection in Body Fluids

Hypericin, pseudohypericin, and hyperforin may be quantitated in plasma as confirmation of usage and to estimate the dosage. These three active substituents have plasma elimination half-lives within a range of 15–60 hours in humans. None of the three has been detected in urine specimens.

Livestock – Forage Poisoning

Clinical Signs: In large doses, St John's wort is poisonous to grazing livestock (cattle, sheep, goats, horses). Behavioural signs of poisoning are general restlessness and skin irritation. Restlessness is often indicated by pawing of the ground, headshaking, head rubbing, and occasional hindlimb weakness with knuckling over, panting, confusion, and depression. Mania and hyperactivity may also result, including running in circles until exhausted.

Observations of thick wort infestations by Australian graziers include the appearance of circular patches giving hillsides a 'crop circle' appearance, it is presumed, from this phenomenon. Animals typically seek shamde and have reduced appetite. Hypersensitivity to water has been noted, and convulsions may occur following a knock to the head. Although general aversion to water is noted, some may seek water for relief.

Severe skin irritation is physically apparent, with reddening of non-pigmented and unprotected areas. This subsequently leads to itch and rubbing, followed by further inflammation, exudation, and scab formation. Lesions and inflammation that occur are said to resemble the conditions seen in foot and mouth disease. Sheep have been observed to have face swelling, dermatitis, and wool falling off due to rubbing. Lactating animals may cease or have reduced milk production; pregnant animals may abort.

Lesions on udders are often apparent. Horses may show signs of anorexia, depression (with a comatose state), dilated pupils, and injected conjunctiva.

Early Diagnosis

Increased respiration and heart rate is typically observed while one of the early signs of St John's wort poisoning is an abnormal increase in body temperature. Affected animals will lose weight, or fail to gain weight; young animals are more affected than old animals. In severe cases death may occur, as a direct result of starvation, or because of secondary disease or septicaemia of lesions. Some affected animals may accidentally drown. Poor performance of suckling lambs (pigmented and non-pigmented) has been noted, suggesting a reduction in the milk production, or the transmission of a toxin in the milk.

Photosensitisation

Most clinical signs in animals are caused by photosensitisation. Plants may induce either primary or secondary photosensitisation:

- primary photosensitisation directly from chemicals contained in ingested plants
- secondary photosensitisation from plant-associated damage to the liver.

Araya and Ford (1981) explored changes in liver function and concluded there was no evidence of *Hypericum*-related effect on the excretory capacity of the liver, or any interference was minimal and temporary. However, evidence of liver damage in blood plasma has been found at high and long rates of dosage.

Photosensitisation causes skin inflammation by a mechanism involving a pigment or photodynamic compound, which when activated by a certain wavelength of light leads to oxidation reactions in vivo. This leads to lesions of tissue, particularly noticeable on and around parts of skin exposed to light.

Lightly covered or poorly pigmented areas are most conspicuous. Removal of affected animals from sunlight results in reduced symptoms of poisoning.

Other Uses

Gardens: *Hypericum perforatum* is a popular groundcover type of ornamental plant used in gardens, including drought tolerant landscaping.

Hypochaeris Radicata

Hypochaeris radicata (sometimes spelled *Hypochoeris radicata*), also known as catsear, flatweed, cat's ear or false dandelion, is a perennial, low-lying edible herb often found in lawns. The plant is native to Europe, but has also been introduced to the Americas, Japan, Australia and New Zealand where it can be an invasive weed. It is listed as a noxious weed in Washington State, USA.

Botany

The leaves, which may grow up to eight inches, are lobed and covered in fine hairs, forming a low-lying rosette around a central taproot. Forked stems carry bright yellow flower heads, and when mature these form seeds attached to windborne "parachutes". All parts of the plant exude a milky sap when cut.

Hypochaeris species are used as food plants by the larvae of some Lepidoptera species including The Shark.

Etymology and Differences from Dandelions

Catsear is derived from the words *cat's ear*, and refers to the shape and fine hair on the leaves resembling that of the ear of a cat.

The plant is also known as *false dandelion*, as it is commonly mistaken for true dandelions. Both plants carry similar flowers which form windborne seeds. However, catsear flowering stems are forked and solid, whereas dandelions possess unforked stems that are hollow. Both plants have a rosette of leaves and a central taproot. The leaves of dandelions are jagged in appearance, whereas those of catsear are more lobe-shaped and hairy. Both plants have similar uses.

Culinary Uses

All parts of the catsear plant are edible; however, the leaves and roots are those most often harvested. The leaves are bland in taste but can be eaten raw in salads, steamed, or used in stir-fries. Older leaves can become tough and fibrous, but younger leaves make for good eating. In contrast to the edible leaves of dandelion, catsear leaves only rarely have some bitterness. In Crete, Greece, the leaves of a variety called pachies or agrioradika are eaten boiled or cooked in steam by the locals.

The root can be roasted and ground to form a coffee substitute.

Toxicity

This species is suspected of causing stringhalt in horses if consumed in excess.

Juglans Nigra

Juglans nigra, the eastern black walnut, a species of flowering tree in the walnut family, Juglandaceae, is native to eastern North America. It grows mostly in riparian zones, from southern Ontario, west to southeast South Dakota, south to Georgia, northern Florida and southwest to central Texas. Isolated wild trees in the upper Ottawa Valley may be an isolated native population or may have derived from planted trees.

The black walnut is a large deciduous tree attaining heights of 30–40 m (98–130 ft). Under forest competition, it develops a tall, clear bole; the open-grown form has a short bole and broad crown.

The bark is grey-black and deeply furrowed. The pith of the twigs contains air spaces. The leaves are alternate, 30–60 cm long, odd-pinnate with 15–23 leaflets, with the largest leaflets located in the centre, 7–10 cm long and 2–3 cm broad.

The male flowers are in drooping catkins 8–10 cm long, the female flowers are terminal, in clusters of two to five, ripening during the autumn into a fruit (nut) with a brownish-green, semifleshy husk and a brown, corrugated nut. The whole fruit, including the husk, falls in October; the seed is relatively small and very hard. The tree tends to crop more heavily in alternate years. Fruiting may begin when the tree is 4–6 years old, however large crops take 20 years. Total lifespan of *J. nigra* is about 130 years.

While its primary native region is the Midwest and east-central United States, the black walnut was introduced into Europe in 1629. It is cultivated there and in North America as a forest tree for its high-quality wood. More nuts are produced by open-grown trees. Black walnut is more resistant to frost than the English or Persian walnut, but thrives best in the warmer regions of fertile, lowland soils with high water tables. It is a light-demanding species.

The wood is used to make furniture, flooring, and rifle stocks, and oil is pressed from the seeds. Nuts are harvested by hand from wild trees. About 65% of the annual wild harvest comes from the U.S. state of Missouri, and the largest processing plant is operated by Hammons Products in Stockton, Missouri. The black walnut nutmeats are used as an ingredient in food, while the hard black walnut shell is used commercially in abrasive cleaning, cosmetics, and oil well drilling and water filtration.

Where the range of *J. nigra* overlaps that of the Texas black walnut *J. microcarpa*, the two species sometimes interbreed, producing populations with characteristics intermediate between the two species.

Uses

Planting: Black walnut plantings can be made to produce timber, nuts, or both timber and nuts. Patented timber-type trees were selected and released from Purdue University in the early 1990s. These trees have been sporadically available from nurseries. Varieties include Purdue #1, which can be used for both timber and nut production, though nut quality is poor compared to varieties selected specifically as nut producers.

Grafted, nut-producing trees are available from several nurseries operating in the U.S. Selections worth considering include Thomas, Neel #1, Thomas Myers, Pounds #2, Stoker, Surprise, Emma K, Sparrow, S127, and McGinnis. Several older varieties, such as Kwik Krop, are still in cultivation; while they make decent nuts, they would not be recommended for commercial planting. A variety index and characteristics guide is available from Missouri Extension.

Pollination requirements should be considered when planting black walnuts. As is typical of many species in Juglandaceae, *Juglans nigra* trees tend to be dichogamous, i.e.. produce pollen first and then pistillate flowers or else produce pistillate flowers and then pollen. An early pollen-producer should be grouped with other varieties that produce pistillate flowers so all varieties benefit from overlap. Cranz, Thomas, and Neel #1 make a good pollination trio. A similar group for more northern climates would be Sparrow, S127, and Mintle.

J. nigra is also grown as a specimen ornamental tree in parks and large gardens, growing to 30 m (98 ft) tall by 20 m (66 ft) broad. It has gained the Royal Horticultural Society's Award of Garden Merit.

Food

Black walnut nuts are shelled commercially in the United States. The nutmeats provide a robust, distinctive, natural flavor and crunch as a food ingredient. Popular uses include ice cream, bakery goods and confections. Consumers include black walnuts in traditional treats, such as cakes, cookies, fudge, and pies, during the fall holiday season. The nuts' nutritional profile leads to uses in other foods, such as salads, fish, pork, chicken, vegetables and pasta dishes.

Nutritionally similar to the milder-tasting English walnut, the black walnut kernel is high in unsaturated fat and protein. An analysis of nut oil from five named *J. nigra* cultivars (Ogden, Sparrow, Baugh, Carter and Thomas) showed the most prevalent fatty acid in *J. nigra* oil is linoleic acid (27.80–33.34 g/100g dry kernel), followed (in the same units) by oleic acid (14.52–24.40), linolenic acid (1.61–3.23), palmitic acid (1.61–2.15), and stearic acid (1.07–1.69). The oil from the cultivar Carter had the highest mol percentage of linoleate (61.6), linolenate (5.97%), and palmitate (3.98%); the oil from the cultivar Baugh had the highest mol percentage of oleate (42.7%); the oil from the cultivar Ogden has the highest mol percentage of stearate (2.98%).

Tapped in spring, the tree yields a sweet sap that can be drunk or concentrated into syrup or sugar.

Nut Processing by Hand

The extraction of the kernel from the fruit of the black walnut is difficult. The thick, hard shell is tightly bound by tall ridges to a thick husk. The husk is best removed when green, as the nuts taste better if it is removed then. Rolling the nut underfoot on a hard surface such as a driveway is a common method; commercial huskers use a car tire rotating against a metal mesh. Some take a thick plywood board and drill a nut-sized hole in it (from one to two inches in diametre) and smash the nut through using a hammer. The nut goes through and the husk remains behind. While the flavor of the *Juglans nigra* kernel is prized, the difficulty in preparing it may account for the wider popularity and availability of the Persian walnut.

Dye

Black walnut drupes contain juglone (5-hydroxy-1,4-naphthoquinone), plumbagin (yellow quinone pigments), and tannin. These compounds cause walnuts to stain cars, sidewalks, porches, and patios, in addition to the hands of anyone attempting to shell them. The brownish-black dye was used by early settlers to dye hair. Extracts of the outer, soft part of the drupe are still used as a natural dye for handicrafts. The tannins present in walnuts act as a mordant, aiding in the dyeing process, and are usable as a dark ink or wood stain.

Wood

Black walnut is highly prized for its dark-coloured, true heartwood. It is heavy and strong, yet easily split and worked. Walnut wood has historically been used for gunstocks, furniture, flooring, paddles, coffins, and a variety of other wood products. Due to its value, forestry officials often are called on to track down walnut poachers; in 2004, DNA testing was used to solve one such poaching case, involving a 55-foot (16-m) tree worth US$2,500. Black walnut has a density of 660 kg per cubic metre (41.2 lb/cubic foot), which makes it less dense than oak.

Pests

Maggots (larvae of *Rhagoletis completa* and *Rhagoletis suavis*) in the husk are common, though more a nuisance than a serious problem for amateurs, who may simply remove the affected husk as soon as infestation is noticed. The maggots develop entirely within the husk, thus the quality of the nutmeat is not affected. However, infestations of maggots are undesirable because they make the husk difficult to remove and are unsightly. Maggots can be serious for commercial walnut growers, who tend to use chemical treatments to prevent damage to the crop. Some organic controls also exist, such as removing and disposing of infested nuts.

The walnut curculio (*Conotrachelus retentus*) grows to 5 mm long as an adult. The adult sucks plant juices through a snout. The eggs are laid in fruits in the spring and summer. Many nuts are lost due to damage from the larvae, which burrow through the nut shell. Codling moth (*Cydia pomonella*) larvae eat walnut kernels, as well as apple and pear seeds.

A disease complex known as thousand cankers disease has been threatening black walnut in several western states. This disease has recently been discovered in Tennessee, and could potentially have devastating effects on the species in the eastern United States.

Toxicity

The roots, nut husks, and leaves secrete a substance into the soil called juglone that is a respiratory inhibitor to some plants. A number of other plants (most notably white birch) are also poisoned by juglone, and should not be planted in close proximity to a black walnut. The plant can cause contact dermatitis in humans. Horses are susceptible to laminitis from exposure to black walnut wood in bedding.

Big Tree

The national champion black walnut is on a residential property in Sauvie Island, Oregon. It is 8 ft 7 in (2.62 m) diametre at breast height and 112 ft (34 m) tall, with a crown spread of 144 feet (44 m).

The largest black walnut in Europe is located in the Castle Park in the city of Sered, Slovakia. It has a circumference of 6.30 metres, height of 25 metres and estimated age of 300 years.

Juniperus Virginiana

Juniperus virginiana (Eastern Red-cedar, Red Cedar, Eastern Juniper, Red Juniper, Pencil Cedar) is a species of juniper native to eastern North America from southeastern Canada to the Gulf of Mexico and east of the Great Plains. Further west it is replaced by the related *Juniperus scopulorum* (Rocky Mountain Juniper) and to the southwest by *Juniperus ashei* (Ashe Juniper).

The Lakota Native American name is Chansha, "redwood" or Hante'. In its native range it is commonly called "cedar" or "red cedar," names rejected by the American Joint Committee on Horticultural Nomenclature as it is a juniper, not a true cedar. However, "Red Cedar" is the most used common name.

Description

Juniperus virginiana is a dense slow-growing coniferous evergreen tree that may never become more than a bush on poor soil, but is ordinarily from 5–20 m/16–66 ft (rarely to 27 m/89 ft) tall, with a short trunk 30–100 cm/12–39 in (rarely 170 cm/67 in) diametre. The oldest tree reported, from Missouri, was 795 years old. The bark is reddish-brown, fibrous, and peels off in narrow strips.

The leaves are of two types; sharp, spreading needle-like juvenile leaves 5–10 cm (2.0–3.9 in) long, and tightly adpressed scale-like adult leaves 2–4 mm (0.079–0.16 in) long; they are arranged in opposite decussate pairs or occasionally whorls of three. The juvenile leaves are found on young plants up to 3 years old, and as scattered shoots on adult trees, usually in shade.

The seed cones are 3–7 mm (0.12–0.28 in) long, berry-like with fleshy scales, dark purple-blue with a white wax cover giving an overall sky-blue colour (though the wax often rubs off); they contain one or two (rarely up to four) seeds, and are mature in 6–8 months from pollination. They are an important winter food for many birds, which disperse the wingless seeds. The pollen cones are 2–3 mm (0.079–0.12 in) long and 1.5 mm (0.059 in) broad, shedding pollen in late winter or early spring. The trees are usually dioecious, with pollen and seed cones on separate trees.

There are Two Varieties, Which Intergrade Where they Meet:

- *Juniperus virginiana* var. *virginiana* is called eastern juniper / redcedar. It is found in eastern North America, from Maine, west to southern Ontario and South Dakota, south to northernmost Florida and southwest into the post oak savannah of east-central Texas. Cones are larger, 4–7 mm; scale leaves are acute at apex and bark is red-brown.

- *Juniperus virginiana* var. *silicicola* (Small) E.Murray (syn. *Sabina silicicola* Small, *Juniperus silicicola* (Small) L.H.Bailey) is known as southern or sand juniper / redcedar. Habitat is along the Atlantic and Gulf coasts from North Carolina, south to central Florida and west to southeast Texas. Cones are smaller, 3–4 mm; scale leaves are blunt at apex and the bark is orange-brown. It is treated by some authors at the lower rank of variety, while others treat it as a distinct species.

Ecology

It is a pioneer invader, which means that it is one of the first trees to repopulate cleared, eroded, or otherwise damaged land. It is unusually long lived among pioneer species, with the potential to live over 850 years. The tree is commonly found in prairies or oak barrens, old pastures, or limestone hills, often along highways and near recent construction sites. It is an alternate host for cedar-apple rust disease, an economically important disease of apples, and some management strategies recommend the removal of *J. virginiana* near apple orchards

In many areas the trees are considered an invasive species, even if native. The fire intolerant *J. virginiana* was previously controlled by periodic wildfires. Low branches near the ground burn and provide a ladder that allows fire to engulf the whole tree. Grasses recover quickly from low severity fires that are characteristic of prairies that kept the trees at bay. With the urbanization of prairies, the fires have been stopped with roads, plowed fields, and other fire breaks, allowing *J. virginiana* and other trees to invade. Trees are destructive to grasslands if left unchecked, and are actively being eliminated by cutting and prescribed burning. The trees also burn very readily, and dense populations were blamed for the rapid spread of wildfires in drought stricken Oklahoma and Texas in 2005 and 2006.

Junipers also benefit from the increased CO_2 levels, unlike the grasses with which they compete. Many grasses are C4 plants that concentrate CO_2 levels in their bundle sheaths to increase the efficiency of RuBisCO, the enzyme responsible for photosynthesis, while Junipers are C3 plants that rely on (and may benefit from) the natural CO_2 concentrations of the environment, although they are less efficient at fixing CO_2 in general.

Damage done by *J. virginiana* includes outcompeting forage species in pastureland. The low branches and wide base occupy a significant portion of land area. The thick foliage blocks out most light, so few plants can live under the canopy. The needles that fall raise the pH of the soil, making it alkaline, which holds nutrients such as phosphorus, making it harder for plants to absorb them. *Juniperus virginiana* has been shown to remove nitrogen from the soil after invading prairie.

It has also been found to reduce carbon stores in the soil. This reduction in soil nutrients also reduces the amount and diversity of microbial activity in the soil.

Cedar waxwings are fond of the "berries" of these junipers. It takes about 12 minutes for their seeds to pass through the birds' guts, and seeds that have been consumed by this bird have levels of germination roughly three times higher than those of seeds the birds did not eat. Many other birds (from bluebirds to turkeys) and many mammals also consume them.

Uses

The fine-grained, soft brittle pinkish- to brownish-red heartwood is fragrant, very light and very durable, even in contact with soil. Because of its rot resistance, the wood is used for fence posts. The aromatic wood is avoided by moths, so it is in demand as lining for clothes chests and closets, often referred to as cedar closets and cedar chests. If correctly prepared, it makes excellent English longbows, flatbows, and Native American sinew-backed bows.

The wood is marketed as "eastern redcedar" or "aromatic cedar". The best portions of the heartwood are one of the few woods good for making pencils, but the supply had diminished sufficiently by the 1940s that it was largely replaced by incense-cedar.

Juniper oil is distilled from the wood, twigs and leaves. The cones are used to flavor gin and as a kidney medicine.

Native American tribes used juniper wood poles to mark out agreed tribal hunting territories. French traders named Baton Rouge, Louisiana (meaning "red stick") from the reddish colour of these poles.

During the Dust Bowl drought of the 1930s, the Prairie States Forest Project encouraged farmers to plant shelterbelts (wind breaks) made of eastern juniper throughout the Great Plains. They grow well under adverse conditions. Both drought tolerant and cold tolerant, they grow well in rocky, sandy, and clay substrate. Competition between trees is minimal, so they can be planted in tightly spaced rows, and the trees still grow to full height, creating a solid windbreak in a short time.

A number of cultivars have been selected for garden planting, including 'Canaertii' (narrow conical; female) 'Corcorcor' (with a dense, erect crown; female), 'Goldspire' (narrow conical with yellow foliage), and 'Kobold' (dwarf). Some cultivars previously listed under this species, notably 'Skyrocket', are actually cultivars of *J. scopulorum*.

In the Missouri and Arkansas Ozarks, eastern juniper is commonly used as a Christmas tree.

Allergen

The pollen is a known allergen, although not as potent as that of the related *Juniperus ashei* (Ashe juniper), which sheds pollen a month earlier. People allergic to one are usually allergic to both. *J. virginiana* sheds pollen as early as late winter and through early spring. Consequently, what begins as an allergy to Ashe juniper in the winter, may extend into spring, since the pollination of the eastern juniper follows after that of the Ashe juniper.

Contact with the leaves or wood can produce a mild skin rash in some individuals.

Kalmia Latifolia

Kalmia latifolia, commonly called mountain-laurel, calico-bush, or spoonwood, is a species of flowering plant in the blueberry family, Ericaceae, that is native to the eastern United States. Its range stretches from southern Maine south to northern Florida, and west to Indiana and Louisiana. Mountain-laurel is the state flower of Connecticut and Pennsylvania. It is the namesake of the city of Laurel, Mississippi (founded 1882).

Growth

It is an evergreen shrub growing to 3–9 m tall. The leaves are 3–12 cm long and 1–4 cm wide. Its flowers are round, ranging from light pink to white, and occurring in clusters. There are several named cultivars today that have darker shades of pink, near red and maroon pigment. It blooms in May and June. All parts of the plant are poisonous. Roots are fibrous and matted. The plant is naturally found on rocky slopes and mountainous forest areas. It thrives in acidic soil, preferring a soil pH in the 4.5 to 5.5 range.

The plant often grows in large thickets, covering great areas of forest floor. In North America it can become tree sized on undeveloped mountains of the Carolinas but is a shrub farther north. The species is a frequent component of oak-heath forests.

Etymology

It is also known as Ivybush, Spoonwood (because native Americans used to make their spoons out of it), Sheep Laurel, Lambkill and Clamoun.

The plant was first recorded in America in 1624, but it was named after Pehr Kalm, who sent samples to Linnaeus in the 18th century.

Cultivation

The plant was originally brought to Europe as an ornamental plant during the 18th century. It is still widely grown for its attractive flowers. Numerous cultivars have been selected with varying flower colour. Many of the cultivars have originated from the Connecticut Experiment Station in Hamden and from the plant breeding of Dr. Richard Jaynes. Jaynes has numerous named varieties that he has created and is considered the world's authority on *Kalmia latifolia*.

This plant and the cultivar 'Pink Charm' have gained the Royal Horticultural Society's Award of Garden Merit.

Wood

Figure: *Wood railing section made with mountain laurel branches*

The wood of the mountain laurel is heavy and strong but brittle, with a close, straight grain. It has never been a viable commercial crop as it does not grow large enough yet it is suitable for wreaths, furniture, bowls and other household items. It was used in the early 19th century in wooden-works clocks. Burls were used for pipe bowls in place of imported briar burls. It can be used for handrails or guard rails, specifically mountain laurel handrail.

Toxicity

Mountain laurel is poisonous to several different animals due to grayanotoxin and arbutin, including horses, goats, cattle, deer, monkeys and humans. The green parts of the plant, flowers, twigs, and pollen are all toxic, including food products made from them, such as honey. Symptoms of toxicity begin to appear about 6 hours following ingestion. Symptoms include irregular or difficulty breathing, anorexia, repeated swallowing, profuse salivation, watering of the eyes and nose, cardiac distress, incoordination, depression, vomiting, frequent defecation, weakness, convulsions, paralysis, coma, and eventually death. Autopsy will show gastrointestinal irritation and hemorrhage.

Bassia Scoparia

Bassia scoparia (synonym *Kochia scoparia*) is an annual shrub native to Eurasia. It has introduced populations in many parts of North America, where it is found in grassland, prairie, and desert shrub ecosystems. Its vernacular names include burningbush, ragweed, summer cypress, fireball, belvedere and Mexican firebrush, Mexican fireweed. It may be planted in almost any climate zone in early spring.

Mexican Firebrush (*B. S. trichophylla*) is a cultivar of *B. scoparia* that turns bright red in the fall. They easily self-seed and can become a weed if not controlled.

Biology

The seed of *Bassia scoparia* is dispersed by wind, water, and especially by the whole plant detaching and tumbling in the wind. The seed does not persist in the soil seed bank, but either germinates or dies within about a year.

Bassia scoparia is a C_4 plant, specifically of the NADP-ME type.

Uses

Uses of *Bassia scoparia* include human food and traditional medicine, forage for livestock, and erosion control.

Tonburi

The seeds of *Bassia scoparia* are eaten as a food garnish called *tonburi* (h0“0v0Š0) (Japanese). Its texture is similar to caviar, and it also is called “land caviar”, “field caviar” and “mountain caviar”. In Japan, tonburi is a delicacy (*chinmi*) of Akita prefecture. After harvesting the seeds are dried. To prepare them, the seeds are boiled and soaked in cold water for about a day, then rubbed by hand to remove the outer skin. The seeds are 1–2 mm in diametre, glossy with a black-green colour.

Tonburi also is used in traditional Chinese medicine. It may prevent metabolic disorders such as hyperlipidemia, hypertension, obesity and atherosclerosis. In a study of mice fed a high-fat diet, an extract of tonburi did limit obesity. *Bassia scoparia* seeds contain momordin Ic, a triterpene saponin.

Forage

The plant is a moderately useful forage for livestock, and a potential forage crop for dry lands. However, its use is limited by toxicity when fed in large quantities. Livestock grazing principally on lush stands of *Bassia scoparia* sometimes experience weight loss, hyperbilirubinemia, photosensitization, and polyuria. When used as the only feed for weeks, *Bassia scoparia* hay may cause toxicity in cattle.

Other

Bassia scoparia is planted for ornament or erosion control. It is a known hyperaccumulator of Chromium, Lead, Mercury, Selenium, Silver, Zinc, and Uranium, , and as such can be used for phytoremediation.

Systematics

The species was first published in 1753 by Carl Linnaeus, who named it *Chenopodium scoparium*. In 1809, it was included into the genus *Kochia* by Heinrich Schrader, and in 1978, into genus *Bassia* by A.J.Scott. Recent phylogenetic research confirmed, that *Kochia* has to be included in *Bassia*.

Chapter 4

Chemical Importance of Poisonous Plants

Cestrum Parqui

Cestrum parqui, commonly known as green cestrum, green poison berry, Chilean cestrum or willow-leaved jessamine, and is sometimes incorrectly referred to as deadly nightshade, is a species of flowering plant in the family Solanaceae that is native to central and South America.

It is an upright, straggly, woody deciduous or semi-evergreen shrub 2–3 metres (6 ft 7 in–9 ft 10 in) tall with one or more brittle green stems. Light green leaves are alternate and shiny green to 12 cm (5 in) long, giving off a foul rubbery smell when crushed. It has sprays of small, fragrant, tubular yellow-green flowers approximately 2.5 cm long on the ends of the stems, flowering from late spring to autumn. These produce clusters of small, black egg shaped berries during summer to autumn.

Green cestrum is highly attractive to birds, and seedlings are often found growing under perching trees, along fence lines and on creek banks. It is also dispersed by water.

Spread by birds, it invades gardens, rural lands and bushland. It has a deep and persistent taproot. This weed is considered a major problem because of its toxicity to livestock (especially cattle) and poultry which eat green cestrum when there is a shortage of other feed. All parts of the plant material, stems, leaves, berries and even partly burnt roots pose a serious threat to livestock. Death is usually rapid and painful. The plant is also known to be toxic to other livestock and humans.

This species is defined as a noxious plant in New South Wales under the Noxious Weeds Act, 1993 and all plants must be destroyed.

In cultivation in the United Kingdom, this plant has gained the Royal Horticultural Society's Award of Garden Merit.

Ericameria Nauseosa

Ericameria nauseosa (formerly *Chrysothamnus nauseosus*), commonly known as Chamisa, rubber rabbitbrush, and gray rabbitbrush, is a shrub of the genus *Ericameria* that grows in the arid regions of western North America. Two subspecies have been described, nauseosa (the gray form with 14 varieties) and consimilis (the green form with 8 varieties).

Rubber rabbitbrush was moved from the genus *Chrysothamnus* to the genus *Ericameria* in a 1993 paper. The findings of a 2003 phylogenetic investigation of *Ericameria* were consistent with the move of the rubber rabbitbrush from *Chrysothamnus* to *Ericameria*. The second edition of the Jepson plant manual and the USDA Germplasm Resources Information Network have adopted the name *Ericameria nauseosa*.

Ericameria nauseosa is a shrubby, 12- to 90-inch (30–230 cm) perennial, producing pungent-smelling, golden-yellow flowers in late summer and early fall. Flower heads are made up of 5 small, yellow, tubular flowers, and occur in umbrella-shaped terminal clusters. The shrub reproduces from seeds and root sprouts. Leaves, depending on the subspecies, are long and narrow to spatula-shaped. Both the flexible (rubbery) stems and the leaves are greenish-gray with a soft felt-like covering.

Cultivation

Rabbitbrush, *Ericameria nauseosa*, has gained popularity as an ornamental xeriscaping shrub in areas where water conservation is important. It thrives in a wide range of coarse, alkaline soils that are common to desert environments. Pruning the shrub back to several inches in early spring, before new growth begins, may help improve the shrub's ornamental value.

Forage

Along with associated species, like big sage and western wheat grass, rubber rabbitbrush is a significant source of food for browsing wildlife on winter ranges. Dense stands of this species often grow on poorly managed rangelands, in disturbed areas along roadways and on abandoned agricultural property.

Possible Commercial Uses

Rubber rabbitbrush was considered as a source of rubber as early as 1904. Several studies have been conducted on the possible use of rubber rabbitbrush as a source of rubber including ones during World Wars I and II, and 1987.

Currently the University of Nevada is conducting research on possible of uses of rubber rabbitbrush for biomaterial and bioenergy uses. One possible commercial use of rubber rabbitbrush would be as a source for hypoallergenic rubber for use in products designed for people with latex allergies.

Radioactivity

Specimens growing in Bayo Canyon, near Los Alamos, New Mexico, exhibit a concentration of radioactive strontium-90 300,000 times higher than a normal plant. Their roots reach into a closed nuclear waste treatment area, mistaking strontium for calcium due to its similar chemical properties.

The radioactive shrubs are "indistinguishable from other shrubs without a Geiger counter."

Cicuta

Cicuta, commonly known as water hemlock, is a small genus of four species of highly poisonous plants in the family Apiaceae. They are perennial herbaceous plants which grow up to 2.5 metres (8.2 ft) tall, having distinctive small green or white flowers arranged in an umbrella shape (umbel).

Plants in this genus may also be referred to as cowbane or poison parsnip. *Cicuta* is native to temperate regions of the Northern Hemisphere, mainly North America and Europe, typically growing in wet meadows, along streambanks and other wet and marshy areas.

These plants bear a close resemblance to other members in the family Apiaceae and may be confused with a number of other edible and poisonous plants. The common name hemlock may also be confused with poison hemlock (*Conium maculatum*).

Water hemlock is considered one of North America's most toxic plants being highly poisonous to humans. Three members of the genus contain a toxin named cicutoxin which causes central nervous system stimulatory effects including seizures following ingestion. Medical treatment of poisoning may include the use of activated charcoal to decrease gastrointestinal absorption of the toxic principle along with supportive care including anticonvulsant drugs such as a benzodiazepine. High doses of anticonvulsant medicine are often required to halt seizure activity and further medical care including intubation and mechanical ventilation may be required.

Description

Cicuta spp. are perennial plants that are all similar in morphology, growing up to a maximum of 2.5 metres (8.2 ft) in height. The stem of the plant is branching, erect, smooth and hollow (except for partitions at the junction of the leaves and stem), sometimes being purple-striped, or mottled (typically only *C. maculata* has the purple stripes or spots). Attached to the base of the stem is a tuberous root with thickened rootstocks.

The rootstocks are multichambered and contain a yellowish oily liquid which turns reddish brown on exposure to air and emits a characteristic smell of raw parsnip.

The alternate leaves are 2 or 3 pinnately compound and may reach 30 centimetres (12 in) to 90 centimetres (35 in) in length. The leaflets are lanceolate, serrate, 5 centimetres (2.0 in) to 10 centimetres (3.9 in) in length, and sharply toothed.

The plant flowers in spring or early summer; the flowers are small with green or white petals clustered in an umbrella shape (umbel) characteristic to this family; the umbel measures 5 centimetres (2.0 in) to 10 centimetres (3.9 in) across. The plants produce a cylindrical fruit which is 4 millimetres (0.16 in) to 6 millimetres (0.24 in) in length. The plant is spread primarily by seeds which are produced in large numbers and are small in size.

Taxonomy

The *Cicuta* genus is one of many genera in the Apiaceae family which is in the order Apiales. The Apiaceae family is also known as Umbelliferae and both of these family names are permitted to be used by the International Code of Botanical Nomenclature.

In Europe, *Cicuta* was not distinguished from the similar genus *Conium* before the year 1500. The first mention of the genus in the United States was in the eighteenth century. Carl Linnaeus formally described three species in 1753.

The type species is *C. virosa*. The genus is now recognized to comprise four species:

Species Name	Common Name
Cicuta bulbifera L.	bulblet-bearing water hemlock, bulbous water hemlock
Cicuta douglasii (DC.) Coult. & Rose	Douglas water hemlock, western water hemlock
Cicuta maculata L.	spotted cowbane, spotted parsley, spotted water hemlock
Cicuta virosa L.	cowbane, Mackenzie's water hemlock, northern water hemlock

Other species names such as *Cicuta bolanderi*, *Cicuta californica*, and *Cicuta curtissii* are older names now recognized to be varieties of the widespread, morphologically variable *Cicuta maculata*. *Cicuta maculata* is now recognized to have four varieties: var. *maculata*, var. *augustifolia*, var. *victorinii*, and var. *bolanderi*. Phylogenetic analysis using the sequences of nuclear ribosomal DNA internal transcribed spacer (ITS) loci was not conclusive but seems to show that *C. bulbifera* and *C. virosa* are monophyletic, while *C. douglasii* may not be. It was also suggested a specimen from California may warrant recognition as a distinct species. Other common names for the genus in general include poison parsnip, beaver poison, wild carrot, wild parsnip, and false parsley.

Similar Species

Members of the family Apiaceae bear close resemblance to each other, and have many characteristics in common. *Cicuta* spp. are often mistaken for edible plants such as sweet flag (*Acorus calamus*), kvanne (*Angelica archangelica*), wild celery (*Apium graveolens*), pignut (*Conopodium majus*), wild carrot (*Daucus carota*), watercress (*Nasturtium* spp.), wild ginseng (*Panax quinquefolius*), wild parsnip (*Pastinaca sativa*), and water parsnip (*Berula* spp.).

One of the more common misidentifications is between water hemlock and water parsnip; both have clusters of small white flowers shaped like umbrellas, and both have the same habitat near the shore line of lakes and rivers. Differences between water parsnip and water hemlock include the water parsnip having leaves only once compound while the water hemlock has leaves which are two or three times compound. Water hemlock also has a large swelling at the stem base which water parsnip lacks. Additionally, water hemlock has bracts at the base of each small flower cluster, not at the base of the main flower head, while water parsnip has both bracts at the base of flowers and also at the main flower head.

Other plants that look similar to *Cicuta* spp. but are not edible are the yarrow, (*Achillea millefolium*, also commonly known as common yarrow, gordaldo, nosebleed plant, old man's pepper, sanguinary, milfoil, soldier's woundwort, thousand-leaf, and thousand-seal), cow Parsnip (*Heracleum lanatum*, also known as *Heracleum maxinium*, Indian celery or pushki) or hogweed (*Heracleum sphondylium*). Features which differentiate them include the yarrow having feathery looking leaves which are pinnately separated into small narrow segments, while cow parsnip and hogweed have large, broad leaves, and an unpleasant odor.

Additionally, there can be confusion between the various water hemlock species and poison hemlock (*Conium maculatum*) as the common name hemlock is applied to both *Cicuta* and *Conium maculatum*. Both are poisonous and can be differentiated by differences in their root structure. Water hemlock has a branched root systems with tubules, while poison hemlock has a single tap root. Another reliable method to identify water hemlock is to examine the leaf veins. Water Hemlock is unique in the Apiaceae family in that it has leaf veins which terminate in the notches between the leaf tips, rather than extend to the tip of the leaf, as is found in the leaf structure of other members of this family.

Distribution and Habitat

Cicuta spp. are found growing across North America and Europe. Typically, they grow in wet habitats usually alongside ponds and streams, in marshes or swamps, or areas that are swampy at least part of the year. Plants can also be found growing in water.

Of the four species, *Cicuta maculata* has the most widespread distribution occurring across the majority of North America. *Cicuta bulbifera* also has a relatively large distribution, found throughout Northern North America. *Cicuta douglasii* is found in the northwest corner of North America, while *Cicuta virosa* is only found in central Europe and in the far north of North America.

Toxicity

All members of *Cicuta* except *C. bulbifera* contain high levels of the poisonous principle cicutoxin, an unsaturated aliphatic alcohol that is structurally closely related to the toxin oenanthotoxin found in the plant hemlock water dropwort. Cicutoxin is present at all stages of growth and in all parts of the plant, but is most concentrated in the roots which appear to be the most toxic in the early spring. Its primary toxic effect is to act as a stimulant in the central nervous system. It is a non-competitive gamma-aminobutyric acid (GABA) receptor antagonist. Cicutoxin acts on the $GABA_A$ receptor causing a block of the chloride channel which results in neuronal depolarization. In the presence of cicutoxin this depolarization continues unabated causing cell overactivity.

The hyperactivity in brain cells results in seizures. Cicutoxin is highly poisonous and water hemlock is considered one of North America's most toxic plants. Ingestion of *Cicuta* can be fatal in humans and there are reports in the medical literature of severe poisoning and death as early as 1670. A number of people have also died following ingestion of the plant in the 20th and 21st century.

The LD_{50} in mice administered cicutoxin by intraperitoneal injection is 48.3 mg per kg body weight (mg/kg); this compares with 5.9 mg/kg for mice given potassium cyanide by intraperitoneal injection, while the LD_{50} for arsenic via intraperitoneal injection in mice is 46.2 mg/kg. The exact toxic dose of plant material in humans is unknown; it is thought ingestion of water hemlock in any quantity can result in poisoning and very small amounts may lead to death.

Poisoning has been reported following children blowing whistles made from the hollow stem of water hemlock plants. Intoxication has also been reported following skin contact with the plant; a case was reported where a family of five people rubbed the plant onto the skin and were poisoned, with two children dying. Livestock have long been the worst affected, leading to the common name "cowbane". Poisoning in livestock is common and typically occurs following ingestion of roots of the plant. In the spring when the ground is soft, grazing animals tend to pull the entire plant out of the ground ingesting both the foiliage and the roots. Roots exposed by ploughing can also be the source of livestock poisonings. Ingestion of plant material may cause death in the animal in as little as 15 minutes.

Symptoms

Upon consumption, both in humans and other species, the symptoms of poisoning are mainly characterized by generalized seizures. The onset of symptoms following ingestion may be as soon as 15 minutes post ingestion. Initial symptoms reported may include nausea, vomiting, abdominal pain, tremors, confusion, weakness, dizziness, and drowsiness; although the rapid onset of seizure activity may be the first sign presented following poisoning. Seizures are usually described as clonic or tonic–clonic. Complications of ongoing seizure activity include increased body temperature, decreases in the pH of the blood (metabolic acidosis), swelling in the brain, blood coagulation disorders, muscle breakdown (rhabdomyolysis), and kidney failure.

Additional neurological symptoms may include hallucinations, delirium, tingling, pricking, or numbness of a person's skin, dilated pupils, and coma. Cardiovascular symptoms include alternating slow or fast heart rate and alternating low and high blood pressure. Other cardiac effects may include ECG abnormalities such as widening of the PR interval, supraventricular tachycardia, and ventricular fibrillation. Symptoms of excess salivation, wheezing, respiratory distress, and absence of breathing have also been reported.

Deaths usually occur from respiratory failure or ventricular fibrillation secondary to ongoing seizure activity; fatalities have occurred within a few hours of ingestion. Poisoned people who recover usually regain consciousness and seizures cease within 24 to 48 hours of poisoning, although seizures may persist for up to 96 hours.

There are occasional long-term effects such as retrograde amnesia of the events leading to intoxication and the intoxication itself. Other ongoing mild effects may include restlessness, muscle weakness, twitching, and anxiety. Complete resolution of symptoms may take a number of days or, in some cases, these ongoing symptoms may persist for months after poisoning.

Diagnosis and Treatment

Water hemlock poisoning is usually diagnosed following a history of plant ingestion and symptoms of abrupt onset of seizures. Laboratory tests to determine the presence of cicutoxin in the blood such as spectrofluorimetry, high pressure liquid chromatography, thin layer chromatography, and mass spectrometry have been used to detect cicutoxin but these tests are not performed routinely in hospital laboratories. If a sample of the plant ingested has been retained, diagnosis can be confirmed by having the plant identified by a botanist.

Initial treatment of poisoning may include gastrointestinal decontamination with activated charcoal. Decontamination is typically only performed if a potentially toxic amount of plant matter has been ingested up to one hour previously and the patient has a normal intact airway or has been intubated. There is no specific antidote for water hemlock poisoning and treatment mainly consists of supportive care.

Treatment may include control of seizures with the administration of a benzodiazepines such as lorazepam or diazepam, or if seizures are refractory to this treatment, a barbiturate such as phenobarbital is administered. The anticonvulsant phenytoin is not recommended as it has not been shown to be effective for seizure control following water hemlock poisoning. Treatment with high doses of benzodiazepines or barbiturates may cause respiratory depression and respiratory support including intubation and mechanical ventilation is required in these patients. Continuous electroencephalography monitoring is recommended in symptomatic patients.

Further treatment for complications of metabolic acidosis, rhabdomyolysis, hyperthermia, or low blood pressure may be required. Metabolic acidosis is treated by administering sodium bicarbonate. Low blood pressure is usually treated with intravenous fluid replacement, but the administration of dopamine or norepinephrine may be required to restore blood pressure. The management of rhabdomyolysis includes ensuring adequate hydration and urinary alkalinization; a complication of rhabdomyolysis is acute renal failure which may require management with hemodialysis. However, hemodialysis, hemoperfusion or other extracorporeal techniques do remove cicutoxin from the blood and are therefore not useful in enhancing elimination.

Ergot

Ergot or ergot fungi refers to a group of fungi of the genus *Claviceps*. The most prominent member of this group is *Claviceps purpurea*. This fungus grows on rye and related plants, and produces alkaloids that can cause ergotism in humans and other mammals who consume grains contaminated with its fruiting structure (called ergot sclerotium). *Claviceps* includes about 50 known species, mostly in the tropical regions. Economically significant species include *C. purpurea* (parasitic on grasses and cereals), *C. fusiformis* (on pearl millet, buffel grass), *C. paspali* (on dallis grass), and *C. africana* (on sorghum). *C. purpurea* most commonly affects outcrossing species such as rye (its most common host), as well as triticale, wheat and barley. It affects oats only rarely.

C. purpurea has at least three races or varieties, which differ in their host specificity:

- G1 — land grasses of open meadows and fields;
- G2 — grasses from moist, forest, and mountain habitats;
- G3 (*C. purpurea* var. *spartinae*) — salt marsh grasses (*Spartina*, *Distichlis*).

Life Cycle

An *ergot kernel* called a *sclerotium* develops when a spore of fungal species of the genus *Claviceps* infects a floret of flowering grass or cereal. The infection process mimics a pollen grain growing into an ovary during fertilization. Infection requires that the fungal spore have access to the stigma, so plants infected by *Claviceps* are mainly outcrossing species with open flowers, such as rye (*Secale cereale*) and ryegrasses (genus *Lolium*).

The proliferating fungal mycelium then destroys the plant ovary and connects with the vascular bundle originally intended for seed nutrition. The first stage of ergot infection manifests itself as a white soft tissue (known as *sphacelia*) producing sugary honeydew, which often drops out of the infected grass florets. This honeydew contains millions of asexual spores (conidia), which insects disperse to other florets. Later, the sphacelia convert into a hard dry sclerotium inside the husk of the floret. At this stage, alkaloids and lipids accumulate in the sclerotium.

Claviceps species from tropic and subtropic regions produce macro- and microconidia in their honeydew. Macroconidia differ in shape and size between the species, whereas microconidia are rather uniform, oval to globose (5x3ìm).

Macroconidia are able to produce secondary conidia. A germ tube emerges from a macroconidium through the surface of a honeydew drop and a secondary conidium of an oval to pearlike shape is formed, to which the contents of the original macroconidium migrates. Secondary conidia form a white, frost-like surface on honeydew drops and spread via the wind. No such process occurs in *Claviceps purpurea*, *Claviceps grohii*, *Claviceps nigricans*, and *Claviceps zizaniae*, all from northern temperate regions.

When a mature sclerotium drops to the ground, the fungus remains dormant until proper conditions trigger its fruiting phase (such as the onset of spring or a rain period). It germinates, forming one or several fruiting bodies with heads and stipe, variously coloured (resembling a tiny mushroom). In the head, threadlike sexual spores form, which are ejected simultaneously when suitable grass hosts are flowering. Ergot infection causes a reduction in the yield and quality of grain and hay, and if livestock eat infected grain or hay it may cause a disease called ergotism. Black and protruding sclerotia of *C. purpurea* are well known. However, many tropical ergots have brown or greyish sclerotia, mimicking the shape of the host seed. For this reason, the infection is often overlooked.

Insects, including flies and moths, carry conidia of *Claviceps* species, but it is unknown whether insects play a role in spreading the fungus from infected to healthy plants.

Effects on Humans and Other Mammals

The ergot sclerotium contains high concentrations (up to 2% of dry mass) of the alkaloid ergotamine, a complex molecule consisting of a tripeptide-derived cyclol-lactam ring connected via amide linkage to a lysergic acid (ergoline) moiety, and other alkaloids of the ergoline group that are biosynthesized by the fungus. Ergot alkaloids have a wide range of biological activities including effects on circulation and neurotransmission.

Ergot alkaloids can be classified into two classes:

1. derivatives of 6,8-dimethylergoline and
2. lysergic acid derivatives.

Ergotism is the name for sometimes severe pathological syndromes affecting humans or other animals that have ingested plant material containing ergot alkaloid, such as ergot-contaminated grains. The Hospital Brothers of St. Anthony (monks) specialized in treating ergotism victims with balms containing tranquilizing and circulation-stimulating plant extracts. The common name for ergotism is “St. Anthony’s Fire”, in reference to monks who cared for victims as well as symptoms, such as severe burning sensations in the limbs. These are caused by effects of ergot alkaloids on the vascular system due to vasoconstriction, sometimes leading to gangrene and loss of limbs due to severely restricted blood circulation.

The neurotropic activities of the ergot alkaloids may also cause hallucinations and attendant irrational behaviour, convulsions, and even death. Other symptoms include strong uterine contractions, nausea, seizures, and unconsciousness. Since the Middle Ages, controlled doses of ergot were used to induce abortions and to stop maternal bleeding after childbirth. Ergot extract has been used in pharmaceutical preparations, including Ergot alkaloids in products such as Cafergot (containing caffeine and ergotamine or ergoline) to treat migraine headaches, and ergometrine, used to induce uterine contractions and to control bleeding after childbirth.

In addition to ergot alkaloids, *Claviceps paspali* also produces tremorgens (paspalitrem) causing “paspalum staggers” in cattle. The fungi of the genera *Penicillium* and *Aspergillus* also produce ergot alkaloids, notably some isolates of the human pathogen *Aspergillus fumigatus*, and have been isolated from plants in the family Convolvulaceae, of which morning glory is best known. Ergot contains no lysergic acid diethylamide (LSD) but instead contains ergotamine, which is used to synthesize lysergic acid, an analog of and precursor for synthesis of LSD. Moreover, ergot sclerotia naturally contain some amounts of lysergic acid.

The January 4, 2007 issue of the *New England Journal of Medicine* includes a paper that documents a British study of more than 11,000 Parkinson's Disease patients.

The study found that two ergot-derived drugs, Pergolide and Cabergoline, commonly used to treat Parkinson's Disease may increase the risk of leaky heart valves by up to 700%.Ergot

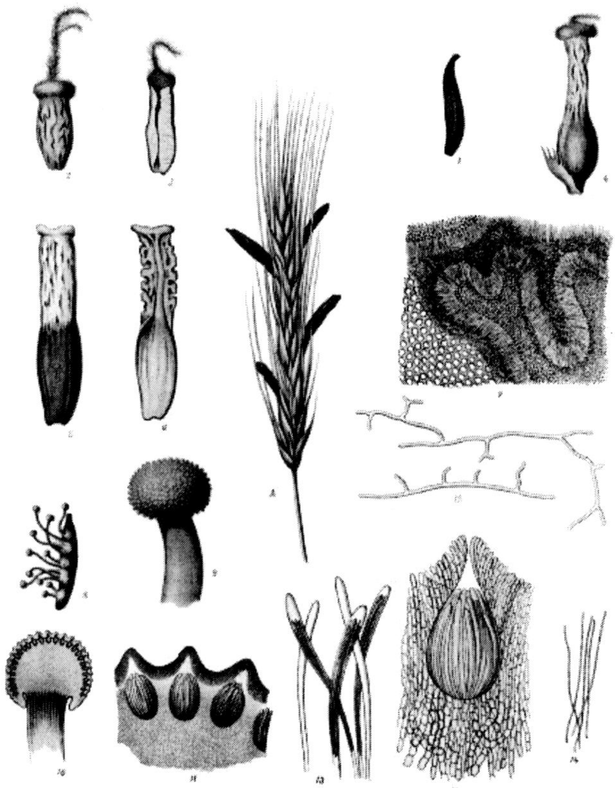

Figure: *Claviceps purpurea*

Scientific Classification	
Kingdom:	Fungi
Division:	Ascomycota
Class:	Sordariomycetes
Order:	Hypocreales
Family:	Clavicipitaceae
Genus:	*Claviceps*

Species

About 50, Including:

Claviceps africana

Claviceps fusiformis

Claviceps paspali

Claviceps purpurea

Claviceps sorghi

Claviceps zizaniae

History

***Figure:** Ergot on wheat spikes*

Human poisoning due to the consumption of rye bread made from ergot-infected grain was common in Europe in the Middle Ages. The epidemic was known as Saint Anthony's fire, or *ignis sacer*, and some historical events, such as the Great Fear in France during the Revolution have been linked to ergot poisoning. Linnda R. Caporael posited in 1976 that the hysterical symptoms of young women that had spurred the Salem witch trials had been the result of consuming ergot-tainted rye. However, Nicholas P. Spanos and Jack Gottlieb, after a review of the historical and medical evidence, later disputed her conclusions. Other authors have likewise cast doubt on ergotism as the cause of the Salem witch trials.

American author John Grigsby contends that the presence of ergot in the stomachs of some of the so-called 'bog-bodies' (Iron Age human remains from peat bogs N E Europe such as Tollund Man) is indicative of use of ergot in ritual drinks in a prehistoric fertility cult akin to the Eleusinian Mysteries cult of ancient Greece.

In his book *Beowulf and Grendel*, he argues that the Anglo-Saxon poem *Beowulf* is based on a memory of the quelling of this fertility cult by followers of Odin. He writes that Beowulf, which he translates as *barley-wolf*, suggests a connection to ergot which in German was known as the 'tooth of the wolf'.

Kykeon, the beverage consumed by participants in the ancient Greek cult of Eleusinian Mysteries, might have been based on hallucinogens from ergot, and lysergic acid diethylamide (LSD) is a potent hallucinogen, which was first synthesized from ergot alkaloids by the Swiss chemist, Albert Hofmann, in 1938.

Claviceps Purpurea

Mankind has known about *Claviceps purpurea* for a long time, and its appearance has been linked to extremely cold winters that were followed by rainy summers.

The sclerotial stage of *C. purpurea* conspicuous on the heads of ryes and other such grains is known as ergot. Favourable temperatures for growth are in the range of 18–30 °C. Temperatures above 37 °C cause rapid germination of conidia. Sunlight has a chromogenic effect on the mycelium, with intense colouration. Cereal mashes and sprouted rye are suitable substrates for growth of the fungus in the laboratory.

Claviceps Africana

Claviceps africana infects sorghum and was first observed in south Texas in 1997. It only infects unfertilized ovaries, so self-pollination and fertilization can decrease the presence of the disease, but male-sterile lines are extremely vulnerable to infection. Symptoms of infection by *C. africana* include the secretion of honeydew (a fluid with high concentrates of sugar and conidia), which attracts insects like flies, beetles, and wasps that feed on it. This in turn contributes to the spread of the fungus to uninfected plants.

C. africana caused false hope that caused a famine in 1903-1906 in Northern Cameroon, West Africa, and also occurs in eastern and southern Africa, especially Zimbabwe and South Africa.

Male sterile sorghums (also referred to as A-lines) are especially susceptible to infection, as first recognized in the 1960s, and massive losses in seed yield have been noted. Infection is associated with cold night temperatures that are below 12 °C occurring two to three weeks before flowering.

Conium

Conium (pron.: /koŠÈnaj.Ym/ or /ÈkoŠniYm/) is a genus of one or two species of highly poisonous perennial herbaceous flowering plants in the family Apiaceae, native to Europe and the Mediterranean region as *Conium maculatum*, and to southern Africa as *Conium chaerophylloides*.

Conium Maculatum

By far the more familiar species is *Conium maculatum* (Hemlock or Poison Hemlock). It is a herbaceous biennial plant that grows between 1.5–2.5 metres (5–8 ft) tall, with a smooth green stem, usually spotted or streaked with red or purple on the lower half of the stem. The leaves are finely divided and lacy, overall triangular in shape, up to 50 centimetres (20 in) long and 40 centimetres (16 in) broad. The flowers are small, white, clustered in umbels up to 10–15 centimetres (4–6 in) across. When crushed, the leaves and root emit a rank, unpleasant odour often compared to that of parsnips.

Name

C. maculatum is known by several common names. As well as the English "Poison hemlock" and the Irish "Devil's Bread", there are also Beaver Poison, Herb Bennet (not to be confused with Herb Robert), Musquash Root, Poison Parsley, Spotted Corobane, and Spotted Hemlock. The seeds are sometimes called Kecksies or Kex.

Figure: *Hemlock seed heads in late summer*

Conium comes from the Greek *konas* (meaning *to whirl*), in reference to vertigo, one of the symptoms of ingesting the plant.

Distribution

C. maculatum is native in temperate regions of Europe, West Asia, as well as North Africa. It has been introduced and naturalised in many other areas, including Asia, North America, Australia, and New Zealand. It is often found on poorly drained soils, particularly near streams, ditches, and other surface water. It also appears on roadsides, edges of cultivated fields, and waste areas. It is considered an Invasive species in twelve U.S. states, including California.

Ecology

C. maculatum grows in damp areas. It is used as a food plant by the larvae of some Lepidoptera species including Silver-ground Carpet. Poison hemlock flourishes in the spring, when most other forage is gone. All plant parts are poisonous but once the plant is dried, the poison is greatly reduced, although not gone completely.

Biochemistry

C. maculatum is a weed known almost worldwide by its toxicity to many domestic animals and to human beings. It is an Apiaceae, characterized by long, hollow stems, reaching up to 2 m (6 ft 7 in) height at maturity, producing a large amount of lush foliage during its vegetative growth.

Its flowers are white, grouped in umbels formed by numerous umbellules. It produces a large number of seeds that allow the plant to form thick stands in modified soils, sometimes encroaching on cultivated fields, to the extent of impeding the growth of any other vegetation inside the *C. maculatum* area of growth.

Eight piperidinic alkaloids have been identified in this species. Two of them, gamma-coniceine and coniine are generally the most abundant and they account for most of the plant's acute and chronic toxicity. These alkaloids are synthesized by the plant from four acetate units from the metabolic pool, forming a polyketoacid which cyclises through an aminotransferase and forms gamma-coniceine as the parent alkaloid via reduction by a NADPH-dependent reductase.

Poison

Conium contains the pyridine alkaloids coniine, N-methylconiine, conhydrine, pseudoconhydrine and gamma-coniceine (or g-coniceïne), which is the precursor of the other hemlock alkaloids.

The most noted of these chemicals is coniine, which has a chemical structure and pharmacological properties similar to nicotine. Coniine disrupts the workings of the central nervous system through action on nicotinic acetylcholine receptors. In high enough concentrations Coniine can be dangerous to humans and livestock.

Due to potency, ingestion can easily result in respiratory collapse and death. Coniine causes death by blocking the neuromuscular junction in a manner similar to curare; this results in an ascending muscular paralysis with eventual paralysis of the respiratory muscles which results in death due to lack of oxygen to the heart and brain. Death can be prevented by artificial ventilation until the effects have worn off 48–72 hours later. For an adult the ingestion of more than 100 mg (.1 grams) of coniine (approximately 6 to 8 fresh leaves, or a smaller dose of the seeds or root) may be fatal.

Isolation of the Alkaloids

Of the total alkaloids of hemlock isolated by the method of Chemnitius and fractionally distilled, the portion boiling up to 190 °C (374 °F) contains most of the coniine, gamma-coniceine and N-methylconiine, while conhydrine and pseudoconhydrine remain in the higher boiling residues. For the separation of coniine from coniceine, Wolffenstein recommends conversion into hydrochlorides. These are dried and extracted with acetone, which dissolves coniceine hydrochloride, leaving the coniine salt, from which the base may then be regenerated. For final purification the coniine is converted into the D-hydrogen tartrate.

It is sometimes necessary to start crystallisation by adding a crystal of the desired salt. Von Braun distills the crude mixed alkaloids until the temperature rises to 190 °C (374 °F), benzoylates the distillate, extracts the tertiary bases by shaking an ethereal solution with dilute acid, pours the concentrated ethereal solution into light petroleum to precipitate most of the benzoyl-ä-aminobutyl propyl ketone formed by the action of benzoyl chloride on coniceine, distills the solvent from the filtrate and collects from the residue the fraction boiling at 200–210 °C (392–410 °F)/16 mmHg (2.1 kPa), which is nearly pure benzoylconiine (bp. 203–204 °C (397–399 °F)/16 mmHg). From this a mixture of D- and L-coniines are obtained by hydrolysis, the former predominating.

Conium Chaerophylloides

The Southern African type *C. chaerophylloides* (resembling Chaerophyllum, Cow parsely) is very similar in appearance to *C.*

maculatum, however the stems of *C. chaerophylloides* are more frequently unspotted. It is found in Zimbabwe, Mozambique and South Africa, in similar habitats to *C. maculatum*. Not all authorities recognize *C. chaerophylloides* as a separate species and the Southern African representatives may be a relatively recent introduction from Europe.

Uses and Effects

Socrates: In ancient Greece, hemlock was used to poison condemned prisoners. The most famous victim of hemlock poisoning is the philosopher Socrates. After being condemned to death for impiety in 399 BC, Socrates was given a potent infusion of the hemlock plant. Plato described Socrates' death in the *Phaedo*:

The man...laid his hands on him and after a while examined his feet and legs, then pinched his foot hard and asked if he felt it. He said "No"; then after that, his thighs; and passing upwards in this way he showed us that he was growing cold and rigid.

And then again he touched him and said that when it reached his heart, he would be gone. The chill had now reached the region about the groin, and uncovering his face, which had been covered, he said – and these were his last words – "Crito, we owe a cock to Asclepius. Pay it and do not neglect it." "That," said Crito, "shall be done; but see if you have anything else to say." To this question he made no reply, but after a little while he moved; the attendant uncovered him; his eyes were fixed. And Crito when he saw it, closed his mouth and eyes.

Although many have questioned whether this is a factual account, careful attention to Plato's words, modern and ancient medicine, and other ancient Greek sources point to the above account being consistent with *Conium* poisoning.

Effects on Animals

C. maculatum is poisonous to animals. In a short time the alkaloids produce a potentially fatal neuromuscular blockage when the respiratory muscles are affected. Acute toxicity, if not lethal, may resolve in the spontaneous recovery of the affected animals provided further exposure is avoided. It has been observed that poisoned animals tend to return to feed on this plant.

Chronic toxicity affects only pregnant animals. When they are poisoned by *C. maculatum* during the fetus's organ formation period, the offspring is born with malformations, mainly palatoschisis and multiple congenital contractures (MCC; frequently described as arthrogryposis).

Chronic toxicity is irreversible and although MCC can be surgically corrected in some cases, most of the malformed animals are lost. Since no specific antidote is available, prevention is the only way to deal with the production losses caused by the plant.

Control with herbicides and grazing with less susceptible animals (such as sheep) have been suggested. *C. maculatum* alkaloids can enter the human food chain via milk and fowl. Such losses may be underestimated, at least in some regions, because of the difficulty in associating malformations with the much earlier maternal poisoning.

Medicinal Uses

Poison hemlock has been used as a sedative and for its antispasmodic properties. It was also used by Greek and Persian physicians for a variety of problems, such as arthritis. It was not always effective, however, as the difference between therapeutic and toxic amounts is very slight. Overdoses can produce paralysis and loss of speech, followed by depression of the respiratory function, and then death.

In Literature and Film

In the 1958 Soviet film *The Penkov Affair* Larisa can be seen making a hemlock "tea" for the attempted murder of rival Tonia.

Hemlock is mentioned by Cordelia in Shakespeare's *King Lear* where she describes her missing father, who has become insane, to the Doctor.

Crowned with rank fumiter and furrow weeds,

With burdocks, hemlock, nettles, cuckoo-flowers,

Darnel, and all the idle weeds that grow

In our sustaining corn.

King Lear, Act V, Scene VI

The plant also appears in John Keats' 1819 poem "Ode to a Nightingale":

"My heart aches, and a drowsy numbness pains

My sense, as though of hemlock I had drunk"

It is also mentioned in the 1986 US action film *Top Gun*, in the bar-scene towards the end of the film, where Charlie finds Maverick drinking and asks jokingly if it's hemlock.

In 2012 noted Bengali director Srijit Bannerjee directed the film *Hemlock Society*, the theme of which is suicide and self-destruction.

In Agatha Christie's "Five Little Pigs" (Poirot Series) "spotted hemlock" is mentioned in Poirot's investigation.

Corydalis

Corydalis (Greek *korydalís* "crested lark") is a genus of about 470 species of annual and perennial herbaceous plants in the Papaveraceae family, native to the temperate Northern Hemisphere and the high mountains of tropical eastern Africa. They are most diverse in China and the Himalayas, with at least 357 species in China.

Corydalis species are used as food plants by the larvae of some Lepidoptera species (butterflies), especially of the genus Mnemosyne.

Corydalis cava and some other tuberous species contain the alkaloid bulbocapnine, which is occasionally used in medicine.

Species

There are about 470 species, including:

- *Corydalis afghanica*
- *Corydalis aitchisonii*
- *Corydalis alpestris*
- *Corydalis ambigua*
- *Corydalis angustifolia*
- *Corydalis aqua-gelidae*
- *Corydalis arctica*
- *Corydalis aurea*
- *Corydalis bracteata*
- *Corydalis buschii*
- *Corydalis caseana*
- *Corydalis caseana ssp. brandegei*
- *Corydalis cashmeriana*
- *Corydalis cava*
- *Corydalis chaerophylla*
- *Corydalis cheilanthifolia*

Ferny Corydalis

- *Corydalis chionophila*
- *Corydalis clavibracteata*
- *Corydalis conorhiza*
- *Corydalis cornuta*

- *Corydalis darwasica*
- *Corydalis diphylla*
- *Corydalis elata*
- *Corydalis emmanuelii*
- *Corydalis flavula*
- *Corydalis flexuosa*
- *Corydalis gigantea*
- *Corydalis glaucescens*
- *Corydalis gortschakovii*
- *Corydalis integra*
- *Corydalis intermedia*
- *Corydalis kushiroensis*
- *Corydalis latiflora*
- *Corydalisleariloba*
- *Corydalis lydica*
- *Corydalis macrocentra*
- *Corydalis marschalliana*
- *Corydalis nariniana*
- *Corydalis nobilis*
- *Corydalis ochotensis*
- *Corydalis ophiocarpa*
- *Corydalis oppositifolia*
- *Corydalis pallida*
- *Corydalis parnassica*
- *Corydalis pinnatibracteata*
- *Corydalis popovii*
- *Corydalis pumila*
- *Corydalis rupestris*
- *Corydalis rutifolia*
- *Corydalis saxicola*
- *Corydalis scouleri*

Scouler's Corydalis

- *Corydalis seisumsiana*
- *Corydalis semenovii*

- *Corydalis shanginii*
- *Corydalis sibirica*
- *Corydalis solida* (fumewort)
- *Corydalis stenantha*
- *Corydalis tomentella*
- *Corydalis trilobipetala*
- *Corydalis turtschaninovii*
- *Corydalis vaginans*
- *Corydalis verticillaris*
- *Corydalis wendelboi*
- *Corydalis wilsonii*
- *Corydalis yanhusuo*

Yan hu so

- *Corydalis zeaensis*

Former Species

Several former *Corydalis* have been moved to new genera:

Pseudofumaria

- *Corydalis lutea = Pseudofumaria lutea*

Yellow Corydalis

- *Corydalis ochroleuca = Pseudofumaria alba*

Capnoides

- *Corydalis sempervirens = Capnoides sempervirens*

Craspedia

Craspedia is a genus of daisies commonly known as billy buttons or woollyheads. They are native to Australia and New Zealand where they grow in a variety of habitats from sea level to the alps. The genus is found in every state of Australia except the Northern Territory. In New Zealand, *Craspedia* is found south from about East Cape in the North Island to Stewart Island. It also occurs on Campbell Island 660 km S of Stewart Island, and the Chatham Islands, 800 km E of East Cape. *Craspedia* are rosette-forming herbs with secondarily compound capitula (glomerules) that are borne on erect, unbranched scapes. The glomerules or flower-heads are hemispherical to spherical (like pom poms) and are formed of a massive aggregation of tiny flowers (florets). Most species are perennial with one species recorded as annual.

Twenty three-species are currently accepted, six from New Zealand and 17 from Australia. Leaves have considerable variation in form, they range in colour from white through to grass green, and are often covered in fine hairs.

Species of *Craspedia* are found in a wide range of habitats from coastal to alpine and are generally plants of open areas, sometimes ruderal. Observations of some Australian species suggest they re-establish well after fire.

In Australia *Craspedia* are commonly found growing in association with forest habitats, whereas in New Zealand they are generally excluded from closed *Nothofagus* forests. *Craspedia* is never seen in New Zealand or Tasmania in the large, widespread and dense populations that are characteristic of mainland Australia, both in forest and on alpine slopes.

Most Australian non-alpine species are found in native grasslands and shrublands associated with *Eucalyptus* forests. Alpine species are characteristic of short grasslands, herbfields and flush habitats of the subalpine and alpine zones of Kosciuszko and Tasmania. Outside the temperate climate zone the only recorded annual, *Craspedia haplorrhiza*, is characteristic of semi-desert steppes, acacia and low eucalypt mallee on the Darling and Riverina floodplains. In New Zealand species exploit an extremely wide range of open habitats ranging from coastal sand dunes to wetland areas, alpine herbfield, fellfield, greywacke rock scree, tall tussock grassland and banks at forest margins. The most unusual habitat is perhaps on the steep, fine greywacke rock screes of the eastern South Island of New Zealand.

Craspedia grows in a wide range of soil types (sands, gravels, clays, earths and loams) derived from different geologies across a broad rainfall gradient. It appears to be intolerant only of extremely infertile and acidic soils. This is illustrated by the absence of *Craspedia* from parts of western Tasmania, about 17% of Tasmania, which are characterised by pre-Cambrian quartzose rocks, which produce infertile sandy acid soils that are wet and dominated by a heath vegetation known as 'buttongrass moorland'.

There are two centres of diversity in *Craspedia*, and both are associated with upland areas. One of these is in the Kosciuzko National Park alpine/subalpine area, where seven species are found. The other centre is a larger area in the north-west of the South Island of New Zealand, which contains several species and about 20 distinct and localised entities.

Cultivation

Craspedia can be propagated by division by cutting a rosette from a clump, but generally seed is a more reliable and rapid method. Seeds will germinate in days if lightly sprinkled on seed raising media and not covered. Plants are generally self fertile. The alpine species are quite fussy about water, needing regular water and excellent drainage. All species prefer cool roots, surrounding the plants with rock, gravel or sand is suitable.

Plants will start growing as a single rosette, and each rosette generally produces only 1 flower stalk. A risky, but effective way to force basal shoots is to let the plant dry out until the leaves wilt, then water the plant. New basal shoots will appear in a few weeks. Tip pruning appears to be ineffective, usually resulting in the death of the plant.

See Also

A closely related genus is *Pycnosorus*, also often called billy buttons. One way to distinguish these genera is to note whether the individual flowers are directly attached to the flower base (*Pycnosorus*) or are attached via little stalks (*Craspedia*).

Crotalaria

Crotalaria is a genus of herbaceous plants and woody shrubs in the Family Fabaceae (Subfamily Faboideae) commonly known as rattlepods. Some 600 or more species of *Crotalaria* are described worldwide, mostly from the tropics; at least 500 species are known from Africa. Some species of *Crotalaria* are grown as ornamentals. The common name rattlepod or *rattlebox* is derived from the fact that the seeds become loose in the pod as they mature, and rattle when the pod is shaken. The name derives from the Greek êñïôáëïí, meaning "castanet", and is the same root as the name for the rattlesnakes (*Crotalus*). *Crotalaria* species are used as food plants by the larvae of some Lepidoptera species including *Endoclita sericeus*, *Etiella zinckenella* and *Utetheisa ornatrix*. The toxic alkaloids produced by some members of this genus are known to be incorporated by *Utetheisia* larvae and used to secure their defense from predators.

- *Crotalaria spectabilis* Roth was introduced to the US from India for green manure. As a legume that supports nitrogen fixing bacteria, it is considered a "soil builder." However, it is also poisonous to cattle (as are many legumes), and has spread rapidly throughout the Southeastern United States where it is now considered an invasive species.

Alkaloid monocrotaline, a pyrrolizidine alkaloid, the main toxic principle of *Crotalaria spectabilis*, is used to induce experimental pulmonary hypertension in laboratory animals.

Larvae of the Ornate moth feed on the plant and re-purpose the poisonous compound as a defense, excreting it when they are threatened by potential predation.

- *Crotalaria longirostrata*, also known as "longbeak rattlebox" or as "chipilín", is a common leafy vegetable in Oaxaca and Central America. It is considered a weed in the United States.
- "Crotalaria pallida" pollen may cause an allergic reaction in humans, including swelling of the eyes and face, a rash on the neck and shoulders, and itching. Symptoms may take up to a week to clear.

Cucumis Myriocarpus

Cucumis myriocarpus, paddy melon or prickly paddy melon is a prostrate or climbing annual herb native to tropical and southern Africa. It has small, round, yellow-green or green-striped fruit with soft spines, small yellow flowers and deeply lobed, light green leaves. The melon occurs in disturbed soil and cleared or bare areas, and thrives on summer moisture.

The fruit and foliage are toxic due to the presence of cucurbitacin. The plant is potentially toxic to horses, sheep, cattle and pigs and has been associated with stock deaths. It has been used by humans as an emetic. There are records of poisoning occurring in humans.

The melon is a weed in Australia and in California, where it may also be known as prickly paddy melon, bitter apple, gooseberry gourd and gooseberry cucumber.

Cucumis myriocarpus (paddy melon) is often confused with *Citrullus lanatus* (the Afghan or camel melon) in Australia, where both species are introduced. *Cucumis myriocarpus* has many small fruit, hidden under the leaves. The fruit are smaller than a golf ball and green in colour, developing to yellow on maturity. The larger melons commonly seen on roadsides in rural Australia are in fact *Citrullus lanatus*, a wild relative of the watermelon.

The confusion is widespread in Australia, such that in common parlance, the term "paddy melon" is understood to mean the larger green/yellow fruit of the *Citrullus lanatus*. Children in rural Australia often use the fruit of this plant as an opportunistic source of entertainment, for example by throwing it at each others' feet.

Paddy and Afghan melons (collectively referred to as wild melons) sometimes occur in mixed infestations in Australia. They vary only slightly in colour and appearance, with the *Citrullus lanatus* leaves slightly more variegated than those of *Cucumis myriocarpus*. Fruit size is the clearest identifying factor. However, management of all wild melons is similar, apart from some variation in sensitivity to glyphosate. Mechanical removal of small infestations is the simplest control method, and cultivation at early stages of growth can be effective. Large infestations are sometimes controlled by the spray-graze method, in which plants are sprayed with a sublethal dose of hormone herbicide to make them more palatable to stock, and the area is heavily grazed three days later. Toxicity is managed by making other feed available.

Both wild melon species have a similar unpleasant odor when broken or disturbed. This makes them less palatable to stock, so poisoning tends only to occur when little other feed is available.

The plant occurs in Spain, where it is naturalized and known by the common names of *habanera* or *sandía habanera*.

Cuscuta

Cuscuta (Dodder) is a genus of about 100-170 species of yellow, orange or red (rarely green) parasitic plants. Formerly treated as the only genus in the family Cuscutaceae, recent genetic research by the Angiosperm Phylogeny Group has shown that it is correctly placed in the morning glory family, Convolvulaceae. The genus is found throughout the temperate to tropical regions of the world, with the greatest species diversity in subtropical and tropical regions; the genus becomes rare in cool temperate climates, with only four species native to northern Europe.

Old folk names include devil's guts, devil's hair, devil's ringlet, goldthread, hailweed, hairweed, hellbine, love vine, pull-down, strangleweed, angel hair, and witch's hair.

Appearance

Dodder can be identified by its thin stems appearing leafless, with the leaves reduced to minute scales. In these respects it closely resembles the similarly parasitic, but unrelated genus Cassytha. From mid-summer to early autumn, the vines can produce small fruit that take the same colour as the vine, and are approximately the size of a common pea. It has very low levels of chlorophyll; some species such as *Cuscuta reflexa* can photosynthesize slightly, while others such as *C. europaea* are entirely dependent on the host plants for nutrition.

Dodder flowers range in colour from white to pink to yellow to cream. Some flower in the early summer, others later, depending on the species. The seeds are minute and produced in large quantities. They have a hard coating, and typically can survive in the soil for 5–10 years, sometimes longer.

Figure: *Cuscuta on acacia tree in Punjab India*

Dodder seeds sprout at or near the surface of the soil. Although dodder germination can occur without a host, it has to reach a green plant quickly and is adapted to grow towards the nearby plants by following chemosensory clues. If a plant is not reached within 5 to 10 days of germination, the dodder seedling will die. Before a host plant is reached, the dodder, as other plants, relies on food reserves in the embryo; the cotyledons, though present, are vestigial.

Parasitism

After a dodder attaches itself to a plant, it wraps itself around it. If the host contains food beneficial to dodder, the dodder produces haustoria that insert themselves into the vascular system of the host. The original root of the dodder in the soil then dies. The dodder can grow and attach itself to multiple plants. In tropical areas it can grow more or less continuously, and may reach high into the canopy of shrubs and trees; in temperate regions it is an annual plant and is restricted to relatively low vegetation that can be reached by new seedlings each spring.

Dodder is parasitic on a very wide variety of plants, including a number of agricultural and horticultural crop species, such as alfalfa, lespedeza, flax, clover, potatoes, chrysanthemum, dahlia, helenium, trumpet vine, ivy and petunias, among others.

Dodder ranges in severity based on its species and the species of the host, the time of attack, and whether any viruses are also present in the host plant. By debilitating the host plant, dodder decreases the ability of plants to resist viral diseases, and dodder can also spread plant diseases from one host to another if it is attached to more than one plant.

Host Finding

A report published in *Science* (Vol.313; Sept. 29, 2006) by Runyon, Mescher, and De Moraes, researchers at Penn State University, demonstrates that dodder use airborne volatile organic compound cues to locate their host plants. Seedlings of *Cuscuta pentagona* exhibit positive growth responses to volatiles released by tomato and other species of host plants. When given a choice between volatiles released by the preferred host tomato and the non-host wheat, the parasite exhibited preferential growth towards the former. Further experiments demonstrated attraction to a number of individual compounds released by host plants and repellance by one compound released by wheat. These results do not rule out the possibility that other cues (e.g., light) may also play a role in host location.

Prevention and Treatment

Figure: *This dodder is engulfing a sage in the Mojave Desert.*

Many countries have laws prohibiting import of dodder seed, requiring crop seeds to be free of dodder seed contamination. Before planting, all clothes should be inspected for dodder seed when moving from an infested area to a non-infested crop. When dealing with an infested area, swift action is necessary.

Recommendations include planting a non-host crop for several years after the infestation, pulling up host crops immediately, particularly before the dodder produces seed, and use of preemergent herbicides such as Dacthal in the spring. Examples of non-host crops include grasses and many other monocotyledons. If dodder is found before it chokes a host plant, it may be simply removed from the soil.

If choking has begun, the host plant must be pruned significantly lower than the dodder, as dodder is versatile and can grow back if present from haustoria.

Selected Species

- *Cuscuta americana*
- *Cuscuta applanata*
- *Cuscuta approximata* Bab. - Alfalfa dodder (Eurasia, Africa)
- *Cuscuta attenuata*
- *Cuscuta australis*
- *Cuscuta boldinghii*
- *Cuscuta brachycalyx*
- *Cuscuta californica* Hook. & Arn. - California dodder (western North America)
- *Cuscuta campestris*
- *Cuscuta cassytoides*
- *Cuscuta ceanothi*
- *Cuscuta cephalanthi*
- *Cuscuta chinensis*
- *Cuscuta compacta*
- *Cuscuta coryli*
- *Cuscuta corylii*
- *Cuscuta cuspidata*
- *Cuscuta decipiens*
- *Cuscuta dentatasquamata*
- *Cuscuta denticulata*

- *Cuscuta epilinum*
- *Cuscuta epithymum* - Clover dodder (Eurasia, Africa)
- *Cuscuta erosa*
- *Cuscuta europaea* - Greater Dodder (Europe)
- *Cuscuta exaltata*
- *Cuscuta fasciculata*
- *Cuscuta gigantea*
- *Cuscuta globulosa*
- *Cuscuta glomerata*
- *Cuscuta gronovii*
- *Cuscuta harperi*
- *Cuscuta howelliana*
- *Cuscuta indecora*
- *Cuscuta indesora*
- *Cuscuta japonica*
- *Cuscuta jepsoni*
- *Cuscuta leptantha*
- *Cuscuta lupuliformis*
- *Cuscuta macrolepis*
- *Cuscuta megalocarpa*
- *Cuscuta monogyna*
- *Cuscuta mitriformis*
- *Cuscuta obtusiflora*
- *Cuscuta odontolepis*
- *Cuscuta pacifica* - Salt Marsh Dodder (British Columbia, Baja California, United States)
- *Cuscuta pentagona* - Golden Dodder (United States)
- *Cuscuta plattensis*
- *Cuscuta polygonorum*
- *Cuscuta potosina*
- *Cuscuta reflexa*
- *Cuscuta rostrata*
- *Cuscuta runyonii*
- *Cuscuta salina* Engelm. - "Salt marsh" dodder (western United States)

- *Cuscuta sandwichiana* Choisy- *Kauna»oa* (Hawai»i)
- *Cuscuta squamata*
- *Cuscuta suaveolens*
- *Cuscuta suksdorfii*
- *Cuscuta tuberculata*
- *Cuscuta umbellata*
- *Cuscuta vivipara*
- *Cuscuta warneri*

Cynoglossum Officinale

Cynoglossum officinale (houndstongue, houndstooth, dog's tongue, gypsy flower, and rats and mice due to its smell) is a herbaceous plant of the family Boraginaceae, found in most parts of Europe, and also North America where it was accidentally introduced.

Growth

It can be either annual or biennial, with reddish-purple flowers blooming between May and September. It lives in wet places, waste land and hedges.

Folklore

The name *houndstongue* comes from the belief that it could ward off dog attacks if a leaf was worn in the shoe.

Herbalism

Herbalists use the plant as a treatment for piles, lung diseases and persistent coughs. Houndstongue ointment is said to cure baldness and be used for sores and ulcers. These uses are not supported by scientific evidence.

In 1725, houndstooth was presented in the family dictionary, *Dictionaire oeconomique*, as part of a cure for madness. In that book, madness was viewed as "a distemper, not only of the understanding, but also of the reason and memory, proceeding from a cold, which drys up everything it meets with that is humid in the brain." To cure madness, *Dictionaire oeconomique* noted:

You must shave the head of the unhappy patient, and after that, apply to it a pidgeon, or a hen quite alive; or else bathe it with some brandy distilled with rosemary, elder, hounds tooth, and the roots of bugloss, or with the oyl of elder flowers: they rub their heads and wash their feet with a decoction of the flowers of camomile, melilot, balm gentle and laurel; they put into their noses the juice of comfrey, with either two or three spoonfuls of honey-water, broth, or white-wine, wherein wormwood and sage are infus'd ; or else they do for five and twenty days together, mix with their broth in the morning, halt a dram of the allies of tortoise, and they put into the pot bugloss, borage, with a pinch of rosemary to season it.

In the 1830s, houndstooth was known in France to be made into an emollient and diuretic for daily use in inflammatory diseases, especially of the urinary organs. To prepare as a diuretic, the houndstooth leaves were mashed, and then boiled in water to extract oils, volatile organic compounds, and other chemical substances.

The mix could be sweetened with liquoriee to create Ptisan of Dog's-grass. After decoction, the herbal tea was taken internally a cupful at a time. In 1834, the Hospital of Paris provided a formula of 2/3 ss—J to Oij of water for houndstooth tea. By the end of the 1830s, doctors in England were using houndstooth as an antiaphrodisiac to combat venereal excesses.

In 1891, the U.S. state of Michigan identified hounds tooth, along with flea-bane, rag weed, burdock, cockle-bur, and stickseed, as some of the worst weeds in the state.

Toxicity

Cynoglossum officinale contains tumorigenic pyrrolizidine alkaloids.

Datura

Datura is a genus of nine species of vespertine flowering plants belonging to the family Solanaceae. They are known as Angel's Trumpets, sometimes sharing that name with the closely related genus *Brugmansia*. They are also sometimes called Moonflowers, one of several plant species to be so. *Datura's* precise and natural distribution is uncertain, owing to its extensive cultivation and naturalization throughout the temperate and tropical regions of the globe. Its distribution within the Americas, however, is most likely restricted to the United States and Mexico, where the highest species diversity occurs.

Some South American plants formerly thought of as *Daturas* are now treated as belonging to the distinct genus *Brugmansia* (*Brugmansia* differs from *Datura* in that it is woody, making shrubs or small trees, and in that it has pendulous flowers, rather than erect ones). Other related genera include *Hyoscyamus* and *Atropa*.

Description

Datura are herbaceous, leafy annuals and short-lived perennials which can reach up to 2 metres in height. The leaves are alternate, 10–20 cm long and 5–18 cm broad, with a lobed or toothed margin. The flowers are erect or spreading (not pendulous like those of the closely allied *Brugmansia*), trumpet-shaped, 5–20 cm long and 4–12 cm broad at the mouth; colours vary from white to yellow, pink, and pale purple. The fruit is a spiny capsule 4–10 cm long and 2–6 cm broad, splitting open when ripe to release the numerous seeds. The seeds disperse freely over pastures, fields and even wasteland locations.

Datura belongs to the classic "witches' weeds," along with deadly nightshade, henbane, and mandrake. Most parts of the plants contain toxic hallucinogens, and datura has a long history of use for causing delirious states and death. It was well known as an essential ingredient of love potions and witches' brews.

The word *datura* comes from the Hindi *dhatûrâ* ("thorn apple"); record of this name dates back to 1662. Nathaniel Hawthorne refers to one type in *The Scarlet Letter* as *apple-Peru*. In Mexico its common name is *toloache*.

The larvae of some Lepidoptera (butterfly and moth) species, including *Hypercompe indecisa*, eat some *datura* species.

Species and Cultivars

It is difficult to classify a datura as to its species, and it often happens that descriptions of new species are accepted prematurely. Later it is found that these "new species" are simply varieties that have evolved due to conditions at a specific location. They usually disappear in a few years. Contributing to the confusion are the facts that various species such as *D. wrightii* and *D. inoxia* are very similar in appearance, and that the variation within a species can be extreme.

For example, *Datura* have the property of being able to change size of plant, size of leaf, and size of flowers, all depending on location. The same species, when growing in a half-shady damp location can develop into a flowering bush half as tall as a person, but when growing in a very dry location will only grow into a thin little plant just higher than the ankles, with tiny flowers and a few miniature leaves.

Today, experts classify only nine species of *Datura*:

- *Datura ceratocaula* Jacq. – Torna Loco
- *Datura discolour* Bernh. – Desert Thorn-apple
- *Datura ferox* L. – Long Spined Thorn-apple

- *Datura inoxia* Mill. – Thorn-apple, Downy Thorn-apple, Indian-apple, Moonflower, Sacred Datura, Toloatzin, Toloache
- *Datura leichhardtii* F.Muell. ex Benth. (syn. *D. pruinosa*) – Leichhardt's Datura
- *Datura metel* L. – Devil's trumpet
- *Datura quercifolia* Kunth – Oak-leaf Thorn-apple
- *Datura stramonium* L. (syn. *D. inermis*) – Jimsonweed, Thorn-apple
- *Datura wrightii* Regel – Sacred Datura, Sacred Thorn-apple

American Brugmansia & Datura Society, Inc. (ABADS), is designated in the 2004 edition of the International Code of Nomenclature for Cultivated Plants as the official International Cultivar Registration Authority for *Datura*. This role was delegated to ABADS by the International Society for Horticultural Science in 2002.

Cultivation

Datura are usually planted annually from the seed produced in the spiny pods, but with care, plants can be overwintered. Most species are suited to being planted outside or in containers. As a rule, they need warm, sunny places and soil that will keep their roots dry. When grown outdoors in good locations, the plants tend to reseed themselves and may become invasive. In containers, they should have porous, aerated potting soil with adequate drainage. The plants are susceptible to fungi in the root area, so organic enrichers such as compost and manure should be avoided.

Toxicity

All *Datura* plants contain tropane alkaloids such as scopolamine, hyoscyamine, and atropine, primarily in their seeds and flowers. Because of the presence of these substances, *Datura* has been used for centuries in some cultures as a poison and as a hallucinogen. There can be a 5:1 toxin variation across plants, and a given plant's toxicity depends on its age, where it is growing, and the local weather conditions. This variation makes *Datura* exceptionally hazardous as a drug.

In traditional cultures, a great deal of experience with and detailed knowledge of *Datura* was critical in order to minimize harm. Many tragic incidents result from modern recreational users ingesting *Datura*. For example, in the 1990s and 2000s, the United States media contained stories of adolescents and young adults dying or becoming seriously ill from intentionally ingesting *Datura*. There are also several reports in the medical literature of deaths from *Datura stramonium* and *Datura ferox* intoxication. Children are especially vulnerable to atropine poisoning, and their prognosis is likely to be fatal. In some parts of Europe and India, *Datura* has been a popular poison for suicide and murder. From 1950–1965, the State Chemical Laboratories in Agra, India investigated 2,778 deaths that were caused by ingesting *Datura*.

The U.S. Centres for Disease Control and Prevention reported accidental poisoning resulting in hospitalization for a family of six who inadvertently ingested *Datura* used as an ingredient in stew. In some places it is prohibited to buy, sell or cultivate *Datura* plants.

Effects of Ingestion

Due to the potent combination of anticholinergic substances it contains, *Datura* intoxication typically produces effects similar to that of an anticholinergic delirium (as contrasted to hallucination): a complete inability to differentiate reality from fantasy; hyperthermia; tachycardia; bizarre, and possibly violent behaviour; and severe mydriasis with resultant painful photophobia that can last several days. Pronounced amnesia is another commonly reported effect.

No other psychoactive substance has received as many severely negative recreational experience reports as has *Datura*. The overwhelming majority of those who describe their use of *Datura* find their experiences extremely unpleasant both mentally and physically and often physically dangerous.

Treatment

Due to their agitated behaviour and confused mental state, victims of *Datura* poisoning are typically hospitalized. Gastric lavage (stomach pumping) and the administration of activated charcoal can be used to reduce the stomach's absorption of the ingested material. The drug physostigmine is used to reverse the effect of the poisons. Benzodiazepines can be given to curb the patient's agitation, and supportive care with oxygen, hydration, and symptomatic treatment is often provided. Observation of the patient is indicated until the symptoms resolve, usually from 24–36 hours after ingestion of the *Datura*.

Delphinium

Delphinium is a genus of about 300 species of perennial flowering plants in the family Ranunculaceae, native throughout the Northern Hemisphere and also on the high mountains of tropical Africa.

The common name "larkspur" is shared between perennial *Delphinium* species and annual species of the genus *Consolida*. Molecular data shows that *Consolida*, as well as another segregate genus, *Aconitella*, are both embedded in *Delphinium*. The name "delphinium" derives from the Latin for "dolphin", referring to the shape of the nectary.

Description

The leaves are deeply lobed with 3 to 7 toothed, pointed lobes in a palmate shape. The main flowering stem is erect, and varies greatly in size between the species, from 10 centimetres in some alpine species, up to 2 metres tall in the larger meadowland species.

The plant is topped with a raceme of many flowers, varying in colour from purple and blue, to red, yellow, or white. In most species each flower consists of five petal-like sepals which grow together to form a hollow pocket with a spur at the end, which gives the plant its name, usually more or less dark blue. Within the sepals are four true petals, small, inconspicuous, and commonly coloured similarly to the sepals. The eponymous long spur of the upper sepal encloses the nectar-containing spurs of the two upper petals.

The seeds are small and often shiny black. The plants flower from late spring to late summer, and are pollinated by butterflies and bumble bees. Despite the toxicity, delphinium species are used as food plants by the larvae of some Lepidoptera species, including the dot moth and small angle shades.

Toxicity

All parts of these plants are considered toxic to humans, causing severe digestive discomfort if ingested, and also skin irritation.

Larkspur, especially tall larkspur, is a significant cause of cattle poisoning on rangelands in the western United States. Larkspur is more common in high-elevation areas, and many ranchers will delay moving cattle onto such ranges until late summer when the toxicity of the plants is reduced. Death is through cardiotoxic and neuromuscular blocking effects, and can occur within a few hours of ingestion.

Species

Species include:

- *Delphinium alabamicum*: Alabama larkspur
- *Delphinium alpestre*: Colourado larkspur
- *Delphinium altissimum*
- *Delphinium andersonii*: Anderson's larkspur
- *Delphinium andesicola*: Chiricahua Mountain larkspur

- *Delphinia antoninum*: Tracy's larkspur
- *Delphinium bakeri*: Baker's delphinium
- *Delphinium barbeyi*: Subalpine larkspur
- *Delphinium basalticum*: Basalt larkspur
- *Delphinium bicolour*: Little larkspur
- *Delphinium brachycentrum*: Northern larkspur
- *Delphinium brownii*
- *Delphinium brunonianum*
- *Delphinium bulleyanum*
- *Delphinium caeruleum*
- *Delphinium californicum*: California larkspur
- *Delphinium cardinale*: Scarlet larkspur
- *Delphinium carolinianum*: Carolina larkspur
- *Delphinium cashmerianum*
- *Delphinium chamissonis*: Chamisso's larkspur
- *Delphinium cheilanthum*
- *Delphinium consolida*
- *Delphinium corymbosum*
- *Delphinium decorum*: Coastal larkspur
- *Delphinium delavayi*
- *Delphinium denudatum*
- *Delphinium depauperatum*: Slim larkspur
- *Delphinium dictyocarpum*
- *Delphinium distichum*: Twospike larkspur
- *Delphinium duhmbergii*
- *Delphinium elatum*: Candle larkspur
- *Delphinium exaltatum* : Tall Larkspur
- *Delphinium fissum*
- *Delphinium formosum*
- *Delphinium geraniifolium*: Clark Valley larkspur
- *Delphinium geyeri*: Geyer's larkspur
- *Delphinium glareosum*: Olympic larkspur
- *Delphinium glaucescens*: Smooth larkspur
- *Delphinium glaucum*: Sierra larkpsur

- *Delphinium gracilentum*: Pine forest larkspur
- *Delphinium grandiflorum*: Siberian larkspur
- *Delphinium gypsophilum*: Pinoche Creek larkspur
- *Delphinium hansenii*: Eldorado larkspur
- *Delphinium hesperium*: Foothill larkspur
- *Delphinium hutchinsoniae*: Monterey larkspur
- *Delphinium hybridum*
- *Delphinium inopinum*: Unexpected larkspur
- *Delphinium leroyi*
- *Delphinium leucophaeum*
- *Delphinium likiangense*
- *Delphinium linarioides*
- *Delphinium lineapetalum*: Thinpetal larkspur
- *Delphinium luteum*: Yellow larkspur
- *Delphinium maackianum*
- *Delphinium macrocentron*
- *Delphinium madrense*: Sierra Madre larkspur
- *Delphinium menziesii*: Menzies' larkspur
- *Delphinium multiplex* : Kittitas larkspur
- *Delphinium muscosum*
- *Delphinium nelsonii*
- *Delphinium newtonianum*: Newton's larkspur
- *Delphinium novomexicanum*: White Mountain larkspur
- *Delphinium nudicaule*: Red larkspur
- *Delphinium nuttallianum*: Twolobe larkspur
- *Delphinium nuttallii*: Upland larkspur
- *Delphinium occidentale*: Subalpine larkspur
- *Delphinium oreganum*
- *Delphinium oxysepalum*
- *Delphinium parishii*: Desert larkspur
- *Delphinium parryi*: San Bernardino larkspur
- *Delphinium patens*: Zigzag larkspur
- *Delphinium pavonaceum*
- *Delphinium peregrinum*

- *Delphinium pictum*
- *Delphinium polycladon*: Mountain marsh larkspur
- *Delphinium przewalskii*
- *Delphinium purpusii*: Kern County larkspur
- *Delphinium pylzowii*
- *Delphinium ramosum*: Mountain larkspur
- *Delphinium recurvatum*: Byron larkspur
- *Delphinium requienii*
- *Delphinium robustum*: Wahatoya Creek larkspur
- *Delphinium roylei*
- *Delphinium sapellonis*: Sapello Canyon larkspur
- *Delphinium scaposum*: Tall mountain larkspur
- *Delphinium scopulorum*: Rocky Mountain larkspur
- *Delphinium semibarbatum*
- *Delphinium speciosum*
- *Delphinium stachydeum*: Spiked larkspur
- *Delphinium staphisagria*
- *Delphinium sutchuense*
- *Delphinium sutherlandii*: Sutherland's larkspur
- *Delphinium tatsienense*
- *Delphinium treleasei*: Glade larkspur
- *Delphinium tricorne*: Dwarf larkspur
- *Delphinium triste*
- *Delphinium trolliifolium*: Columbian larkspur
- *Delphinium uliginosum*: Swamp larkspur
- *Delphinium umbraculorum*: Umbrella larkspur
- *Delphinium variegatum*: Royal larkspur
- *Delphinium verdunense*
- *Delphinium vestitum*
- *Delphinium villosum*
- *Delphinium virescens*
- *Delphinium viridescens*: Wenatchee larkspur
- *Delphinium viride*
- *Delphinium wootonii*: Organ Mountain larkspur

- *Delphinium xantholeucum*: Yellow-white larkspur
- *Delphinium yunnanense*
- *Delphinium zalil*: Zalil

Cultivation

Various *Delphinium* species are cultivated by nurseries as ornamental plants, for traditional and native plant gardens. However, the numerous hybrids and cultivars are primarily used as garden plants. Delphiniums can attract butterflies and other pollinators.

Hybrids and Cultivars

Most Delphinium hybrids and cultivars are derived from *Delphinium elatum* (syn: *Delphinium* × *elatum*), native to the Alps. Hybridisation was developed in the 19th-century, lead by Victor Lemoine in France. Other hybrid crosses have included *Delphinium bruninianum* from the Himalayas, *Delphinium cardinale* from California, *Delphinium cheilanthum* from Siberia, and *Delphinium formosum* from Armenia.

Numerous cultivars have been selected as garden plants, and for cut flowers and floristry. They are available in shades of white, pink, purple, and blue. The blooming plant is also used in displays and specialist competitions at flower and garden shows, such as the Chelsea Flower Show.

The Pacific Giant Hybrid Delphiniums are a group with individual single-colour cultivar names, developed by Reinelt in the United States. They typically grow to 4–6 feet (1.2–1.8 m) tall on long stems, by 2–3 feet (0.61–0.91 m) wide. They reportedly can tolerate deer. Millenium Delphinium hybrids, bred by Dowdeswell's Ltd in New Zealand, are reportedly better in warmer climates than the Pacific Hybrids. Flower colours in shades of red, orange, and pink were hybridized from *Delphinium cardinale*, a California native species, by Americans Reinelt and Samuelson.

Selected Cultivars

The following *Delphinium* cultivars have received the Award of Garden Merit from the British Royal Horticultural Society:

- 'Atholl'
- 'Atlantis' (Belladonna group)
- 'Blue Dawn'
- 'Blue Nile'

- 'Bruce'
- 'Can-can'
- 'Cassius'
- 'Cherub'
- 'Claire'
- 'Clifford Sky'
- 'Conspicuous'
- 'Constance Rivett'
- 'David Mannion'
- 'Elizabeth Cook'
- 'Emily Hawkins'
- 'Faust'
- 'Fenella'
- 'Galileo'
- 'Gillian Dallas'
- 'Giotto'
- 'Langdon's Pandora'
- 'Langdon's Royal Flush'
- 'Lilian Bassett'
- 'Loch Leven'
- 'Lord Butler'
- 'Lucia Sahin'
- 'Michael Ayres'
- 'Min'
- 'Olive Poppleton'
- 'Oliver'
- 'Our Deb'
- 'Pink Sensation' (*D.* x *ruysii*)
- 'Rosemary Brock'
- 'Sandpiper'
- 'Spindrift'
- 'Summerfield Miranda'
- 'Sungleam'
- 'Sunkissed'

- 'Tiddles'
- 'Walton Gemstone'
- 'Völkerfrieden' (Belladonna group)

Other Uses

The juice of the flowers, mixed with alum, gives a blue ink.

A *Delphinium* species was one of the herbs used on the feast of St. John, and as such, warded off lightning. In Transylvania, one was used to keep witches from the stables.

Medicinal

All parts of the plant contain various diterpenoid alkaloids, typified by methyllycaconitine, so are very poisonous. In small amounts, extracts of the plant have been used in herbal medicine.

Delphinium is used to treat intestinal worms, fluid retention, poor appetite, and trouble sleeping (insomnia). It is also used as a sedative to cause relaxation. Gerard's *Herbal* reported that drinking the seed of larkspur was thought to help against the stings of scorpions, and that other poisonous animals could not move when covered by the herb, but that he did not believe this himself. Mrs. Grieve reports that the seeds can be used against parasites, especially lice and their nits in the hair. A tincture is used against asthma and dropsy.

The plant was connected to Saint Odile of Alsace, a patroness of good eyesight, and so in popular medicine was used

Dendrocnide Moroides

Dendrocnide moroides, also known as the gympie gympie, moonlighter, or stinger, is a large shrub native to rainforest areas in northeastern Australia, the Moluccas and Indonesia and in Kodagu, India . It is best known for stinging hairs which cover the whole plant and deliver a potent neurotoxin when touched. It is the most virulent species of stinging tree.

D. moroides usually grows as a single-stemmed plant reaching 1–2 metres in height. It has large, heart-shaped leaves about 12–22 cm long and 11–18 cm wide, with finely toothed margins.

Ecology

The species is unique in the *Dendrocnide* genus in having bisexual inflorescences in which the few male flowers are surrounded by female flowers. The flowers are small, and once pollinated, the stalk swells to form the fruit. Fruits are juicy and mulberry-like and are bright pink to purple. Each fruit contains a single seed which is on the outside of the fruit. The species is an early coloniser in rainforest gaps; seeds germinate in full sunlight after soil disturbance. Although relatively common in Queensland, the species is uncommon in its southern-most range, and is listed as an endangered species in New South Wales.

The giant stinging tree and the shining-leaved stinging tree are other large nettles occurring in Australia.

Toxicity

Contact with the leaves or twigs causes the hollow, silica-tipped hairs to penetrate the skin. The sting causes a painful stinging sensation which can last for days or even months, and the injured area becomes covered with small, red spots joining together to form a red, swollen mass. The sting is known to be potent enough to kill humans, and it can also kill dogs and horses. However, the sting does not stop several small marsupial species, including the red-legged pademelon, insects and birds from eating the leaves. Moroidin, a bicyclic octapeptide containing an unusual C-N linkage between tryptophan and histidine, was first isolated from the leaves and stalks of *Dendrocnide moroides*, and subsequently shown to be the principal compound responsible for the long duration of the stings.

Treatment

The recommended treatment for skin exposure to the hairs is by pulling them out with a wax hair removal strip or rubbing human hair to the affected part (practised in Kodagu India) .

Descurainia Pinnata

Descurainia pinnata is a species of flowering plant in the mustard family known by the common name western tansymustard. It is native to North America, where it is widespread and found in varied habitats. It is especially successful in deserts. It is a hardy plant which easily becomes weedy, and can spring up in disturbed, barren sites with bad soil. This is a hairy, heavily branched, mustardlike annual which is quite variable in appearance.

There are several subspecies which vary from each other and individuals within a subspecies may look different depending on the climate they endure. This may be a clumping thicket or a tall, erect mustard. It generally does not exceed 70 centimetres in height. It has highly lobed or divided leaves with pointed, toothed lobes or leaflets. At the tips of the stem branches are tiny yellow flowers. The fruit is a silique one half to two centimetres long upon a threadlike pedicel. This plant reproduces only from seed. This tansymustard is toxic to grazing animals in large quantities due to nitrates and thiocyanates; however, it is a nutritious in smaller amounts. The flowers are attractive to butterflies. The seeds are said to taste somewhat like black mustard and were utilized as food by Native American peoples such as the Navajo.

Digitalis

Digitalis is a genus of about 20 species of herbaceous perennials, shrubs, and biennials commonly called foxgloves. This genus was traditionally placed in the figwort family Scrophulariaceae, but recent phylogenetic research has placed it in the much enlarged family Plantaginaceae. This genus is native to western and southwestern Europe, western and central Asia, Australasia and northwestern Africa.

The scientific name means "finger-like" and refers to the ease with which a flower of *Digitalis purpurea* can be fitted over a human fingertip.

The flowers are produced on a tall spike, are tubular, and vary in colour with species, from purple to pink, white, and yellow. The best-known species is the common foxglove, *Digitalis purpurea.* This biennial plant is often grown as an ornamental plant due to its vivid flowers which range in colour from various purple tints through various shades of light gray, and to purely white. The flowers can also possess various marks and spottings.

The first year of growth of the common foxglove produces only the stem with its long, basal leaves. During the second year of the plant's life, a long, leafy stem from 50 to 255 centimetres tall grows atop the roots of healthy plants. The larvae of the insect the "foxglove pug" consume the flowers of the common foxglove for food. Other species of Lepidoptera eat the leaves, including lesser yellow underwing.

The term digitalis is also used for drug preparations that contain cardiac glycosides, particularly one called digoxin, extracted from various plants of this genus.

Etymology

Many suggestions for the derivation of the name "foxglove" have been proffered. According to the 19th-century book, *English Botany, Or, Coloured Figures of British Plants*:

Dr. Prior, whose authority is great in the origin of popular names, says "It seems probably that the name was in the first place, *foxes' glew*, or music, in reference to the favourite instrument of an earlier time, a ring of bells hung on an arched support, the *tintinnabulum*"... we cannot quite agree with Dr. Prior for it seems quite probable that the shape of the flowers suggested the idea of a glove, and that associated with the name of the botanist *Fuchs*, who first gave it a botanical name, may have been easily corrupted into *foxglove*.

It happens, moreover, the name *foxglove* is a very ancient one and exists in a list of plants as old as the time of Edward III. The "folks" of our ancestors were the fairies and nothing is more likely than that the pretty coloured bells of the plant would be designated "*folksgloves*," afterwards, "foxglove."

In Wales it is declared to be a favourite lurking-place of the fairies, who are said to occasion a snapping sound when children, holding one end of the digitalis bell, suddenly strike the other on the hand to hear the clap of fairy thunder, with which the indignant fairy makes her escape from her injured retreat. In south of Scotland it is called "bloody fingers" more northward, "deadman's bells" whilst in Wales it is known as "fairy-folks-fingers" or "lambs-tongue-leaves".

Habitat

Digitalis species thrive in acidic soils, in partial sunlight to deep shade, in a range of habitats, including open woods, woodland clearings, moorland and heath margins, sea-cliffs, rocky mountain slopes and hedge banks. It is commonly found on sites where the ground has been disturbed, such as recently cleared woodland, or where the vegetation has been burnt.

Medicinal Uses

Cardiac: A group of medicines extracted from foxglove plants are called *Digitalin*. The use of *D. purpurea* extract containing cardiac glycosides for the treatment of heart conditions was first described in the English-speaking medical literature by William Withering, in 1785, which is considered the beginning of modern therapeutics. It is used to increase cardiac contractility (it is a positive inotrope) and as an antiarrhythmic agent to control the heart rate, particularly in the irregular (and often fast) atrial fibrillation.

Digitalis is hence often prescribed for patients in atrial fibrillation, especially if they have been diagnosed with congestive heart failure. Digoxin was approved for heart failure in 1998 under current regulations by the Food and Drug Administration on the basis of prospective, randomized study and clinical trials. It was also approved for the control of ventricular response rate for patients with atrial fibrillation. American College of Cardiology/American Heart Association guidelines recommend digoxin for symptomatic chronic heart failure for patients with reduced systolic function, preservation of systolic function, and/or rate control for atrial fibrillation with a rapid ventricular response. Heart Failure Society of America guidelines for heart failure provide similar recommendations.

Despite its relatively recent approval by the Food and Drug Administration and the guideline recommendations, the therapeutic use of digoxin is declining in patients with heart failure—likely the result of several factors. Digoxin has not been promoted by the pharmaceutical industry and has received little attention at national and international meetings—possibly due to the development and promotion of other, newly patented therapies for heart failure. Safety concerns regarding a proposed link between digoxin therapy and increased mortality in women may also be contributing to the decline in therapeutic use of digoxin.

Variations

A group of pharmacologically active compounds are extracted mostly from the leaves of the second year's growth, and in pure form are referred to by common chemical names, such as digitoxin or digoxin, or by brand names such as Crystodigin and Lanoxin, respectively. The two drugs differ in that digoxin has an additional hydroxyl group at the C-3 position on the B-ring (adjacent to the pentane). Both molecules include a lactone and a triple-repeating sugar called a glycoside.

Mechanism of Action

Digitalis works by inhibiting sodium-potassium ATPase. This results in an increased intracellular concentration of sodium ion and thus a decreased concentration gradient across the cell membrane. This increase in intracellular sodium makes the actions of the Na/Ca exchange pump unfavourable, thereby decreasing sodium influx and calcium efflux via this pump. This gradient-induced inhibition results in a higher cytoplasmic calcium concentration.

The increased cytosolic calcium ion concentration results in increased calcium ion storage in the sarcoplasmic reticulum. Upon action potential (cardiac contraction), more calcium is released from the sarcoplasmic reticulum and this gives a positive inotropic effect (higher contractility). Digitalis also has a vagal effect on the parasympathetic nervous system, and as such is used in re-entrant cardiac arrhythmias and to slow the ventricular rate during atrial fibrillation. The dependence on the vagal effect means digitalis is not effective when a patient has a high sympathetic nervous system drive, which is the case with acutely ill persons, and also during exercise.

Toxicity

Digitalis toxicity (*Digitalis intoxication*) results from an overdose of digitalis and causes anorexia, nausea, vomiting and diarrhoea, as well as sometimes resulting in xanthopsia (jaundiced or yellow vision) and the appearance of blurred outlines (halos). Bradycardia also occurs. Because a frequent side effect of digitalis is reduction of appetite, some individuals have used the drug as a weight-loss aid.

Digitalis is an example of a drug derived from a plant that was formerly used by folklorists and herbalists; herbalists have largely abandoned its use because of its narrow therapeutic index and the difficulty of determining the amount of active drug in herbal preparations. Once the usefulness of digitalis in regulating the human pulse was understood, it was employed for a variety of purposes, including the treatment of epilepsy and other seizure disorders, which are now considered to be inappropriate treatments.

Depending on the species, the digitalis plant may contain several deadly physiological and chemically related cardiac and steroidal glycosides. Thus, the digitalis plants have earned several, more sinister, names: dead man's bells and witch's gloves.

The entire plant is toxic (including the roots and seeds). Mortality is rare, but case reports do exist. Most plant exposures occur in children younger than six years and are usually unintentional and without associated significant toxicity. More serious toxicity occurs with intentional ingestions by adolescents and adults. Early symptoms of ingestion include nausea, vomiting, diarrhoea, abdominal pain, wild hallucinations, delirium, and severe headache. Depending on the severity of the toxicosis, the victim may later suffer irregular and slow pulse, tremors, various cerebral disturbances, especially of a visual nature (unusual colour visions with objects appearing yellowish to green, and blue halos around lights), convulsions, and deadly disturbances of the heart. For a case description.

In some instances, people have confused digitalis with the relatively harmless comfrey (*Symphytum*) plant, which is often brewed into a tea, with fatal consequences. Other fatal accidents involve children drinking the water in a vase containing digitalis plants. Drying does not reduce the toxicity of the plant. The plant is toxic to animals, including all classes of livestock and poultry, as well as felines and canines.

Digitalis poisoning can cause heart block and either bradycardia (decreased heart rate) or tachycardia (increased heart rate), depending on the dose and the condition of one's heart. Notably, the electric cardioversion (to "shock" the heart) is generally not indicated in ventricular fibrillation in digitalis toxicity, as it can increase the dysrhythmia.

Also, the classic drug of choice for ventricular fibrillation in emergency setting, amiodarone, can worsen the dysrhythmia caused by digitalis, therefore, the second-choice drug lidocaine is more commonly used.

Digoxigenin

Digoxigenin (DIG) is a steroid found exclusively in the flowers and leaves of the plants *D. purpurea* and *D. lanata*. It is used as a molecular probe to detect DNA or RNA. It can easily be attached to nucleotides by chemical modifications. DIG molecules are often linked to uridine nucleotides; DIG-labelled uridine (DIG-U) can then be incorporated into RNA probes via *in vitro* transcription. Once hybridisation occurs *in situ*, RNA probes with the incorporated DIG-U can be detected with anti-DIG antibodies that are conjugated to alkaline phosphatase. To reveal the hybridised transcripts, alkaline phosphatase can be reacted with a chromogen to produce a coloured precipitate.

Dryopteris Filix-Mas

Dryopteris filix-mas (male fern) is a common fern of the temperate Northern Hemisphere, native to much of Europe, Asia, and North America. It favours damp shaded areas in the understory of woodlands, but also shady places on hedge-banks, rocks, and screes.

It is much less abundant in North America than in Europe. The plant is sometimes referred to in ancient literature as worm fern.

Description

The semi-evergreen leaves have an upright habit and reach a maximum length of 150 cm (59 in), with a single crown on each rootstock. The bipinnate leaves consist of 20-35 pinnae on each side of the rachis. The leaves taper at both ends, with the basal pinnae about half the length of the middle pinnae. The pinules are rather blunt and equally lobed all around. The stalks are covered with orange-brown scales. On the abaxial surface of the mature blade 5 to 6 sori develop in two rows. When the spores ripen in August to November, the indusium starts to shrivel, leading to the release of the spores.

This species hybridises easily with *Dryopteris affinis* (scaly male fern) and *Dryopteris oreades* (mountain male fern).

Cultivation and Uses

Numerous cultivars have been selected for garden use. The species and the cultivar 'Cristata' have both gained the Royal Horticultural Society's Award of Garden Merit.

The root was used, until recent times, as an anthelmintic to expel tapeworms, but has been replaced by less toxic and more effective drugs. The anthelmintic activity has been claimed to be due to flavaspidic acid, a phloroglucinol derivative.

Duboisia

Duboisia (commonly called Corkwood Tree) is a genus of small perennial shrubs to trees about 14 m tall, with extremely light wood and a thick corky bark. There are four species; all occur in Australia, and one also occurs in New Caledonia.

The alternate, glabrous leaves are narrow and elliptical. The inflorescence is an open cymose panicle of apically small white flowers, sometimes with a purple or mauve striped tube. They flower profusely in spring. The fruit is a small, globular, black, juicy berry.

The nicotine, scopolamine and hyoscyamine-containing leaves of *Duboisia* are the active component of the drug *pituri*, used by indigenous peoples of central Australia for its stimulant, euphoric, antispasmodic and analgesic effects.

For example, smoke from the burning leaves is inhaled at ceremonies, such as male initiation rites, including circumcision, in part because of its anaesthetic properties and ability to induce altered states of consciousness.

The dried leaves of *Duboisia hopwoodii* is also mixed with ash from *Acacia* species and chewed. One paleontologist, Dr Gavin Young, named the fossil agnathan *Pituriaspis doylei* after pituri, as he thought he might be hallucinating upon viewing the fossil fish's bizarre form. Prior to European settlement in Australia, *pituri* was exported along trade routes and used by indigenous peoples far from central Australia, including Torres Strait Islanders.

The leaves of *Duboisia leichhardtii* and *Duboisia myoporoides* also contain scopolamine and hyoscyamine, along with some other pharmaceutically important alkaloids. A derivative of scopolamine is the drug butylscopolamine, a potent peripherally acting antispasmodic. These trees are commercially grown for the pharmaceutical industry.

Echium Plantagineum

Echium plantagineum, commonly known as Purple Viper's Bugloss or Paterson's Curse, is a species of *Echium*, native to western and southern Europe (from southern England south to Iberia and east to the Crimea), northern Africa, and southwestern Asia (east to Georgia). It has also been introduced to Australia, South Africa and United States, where it is an invasive weed.

Due to a high concentration of pyrrolizidine alkaloids in the shoot, it is poisonous to grazing livestock, especially those with a simple digestive system like horses. The toxins are cumulative in the liver, and death results from too much Paterson's curse in the diet.

Description

Echium plantagineum is a winter annual plant growing to 20–60 cm tall, with rough, hairy, lanceolate leaves up to 14 cm long. The flowers are purple, 15–20 mm long, with all the stamens protruding, and borne on a branched spike.

Invasive Species

E. plantagineum has become an invasive species in Australia, where it is also known as Paterson's Curse, Patterson's Curse or Salvation Jane (particularly in South Australia). Other names are *Blueweed, Lady Campbell Weed* or *Riverina Bluebell.*

In the United States the species has become naturalised in parts of California, Oregon and some eastern states. In Oregon it has been declared a noxious weed.

Medical Research

Seeds of the *E. plantagineum* plant were found in a study funded by the National Centre for Complementary and Alternative Medicine (part of the National Institutes of Health) to lower triglycerides. Researchers at Wake Forest University and the Harvard Centre for Botanical Lipids fed mice a diet supplemented with echium oil and found that it had benefit similar to fish oil in lowering triglyceride levels in blood plasma and the liver.

Cosmetics

Echium oil contains high levels of alpha linolenic acid (ALA), gamma linolenic acid (GLA) and stearidonic acid (SDA), making it valuable in cosmetic and skin care applications, with further potential as a functional food, as an alternative to fish oils.

Toxicity

Echium plantagineum contains pyrrolizidine alkaloids and is poisonous. When eaten in large quantities, it cause reduced livestock weight or even (in severe cases) death. Paterson's Curse can kill horses and irritate the udders of dairy cows and the skin of humans. After the 2003 Canberra bushfires over 40 recorded horses were put down after eating the weed. Because the pyrrolizidine alkaloids can also be found in the nectar of Paterson's curse the honey made from it should be blended with other honey to avoid the toxins entering the human food chain in high concentrations. The alkaloid levels in one teaspoon of salvation jane honey are generally above the EU recommended weekly maximum human consumption for these chemicals.

Equisetum

Equisetum (pron.: /Ì[kwhÈsiÐtYm/; horsetail, snake grass, puzzlegrass) is the only living genus in Equisetaceae, a family of vascular plants that reproduce by spores rather than seeds.

Equisetum is a "living fossil" as it is the only living genus of the entire class Equisetopsida, which for over one hundred million years was much more diverse and dominated the understory of late Paleozoic forests. Some Equisetopsida were large trees reaching to 30 metres tall. The genus *Calamites* of the family Calamitaceae, for example, is abundant in coal deposits from the Carboniferous period.

A superficially similar but entirely unrelated flowering plant genus, mare's tail (*Hippuris*), is occasionally misidentified as "horsetail".

It has been suggested that the pattern of spacing of nodes in horsetails, wherein those towards the apex of the shoot are increasingly close together, inspired John Napier to discover logarithms.

Etymology

The name "horsetail", often used for the entire group, arose because the branched species somewhat resemble a horse's tail. Similarly, the scientific name *Equisetum* derives from the Latin *equus* ("horse") + *seta* ("bristle").

Other names include candock for branching individuals, and scouring-rush for unbranched or sparsely branched individuals. The latter name refers to the plants' rush-like appearance, and to the fact that the stems are coated with abrasive silicates, making them useful for scouring (cleaning) metal items such as cooking pots or drinking mugs, particularly those made of tin. In German, the corresponding name is *Zinnkraut* ("tin-herb"). Rough horsetail *E. hyemale* is still boiled and then dried in Japan, to be used for the final polishing process on woodcraft to produce a smoother finish than any sandpaper.

Distribution, Ecology and Uses

The genus *Equisetum* is near-cosmopolitan, being absent only from Antarctica. They are perennial plants, either herbaceous and dying back in winter as most temperate species, or evergreen as most tropical species and the temperate species rough horsetail (*E. hyemale*), branched horsetail (*E. ramosissimum*), dwarf horsetail (*E. scirpoides*) and variegated horsetail (*E. variegatum*). They typically grow 0.2-1.5 m tall, though the "giant horsetails" are recorded to grow as high as 2.5 m (northern giant horsetail, *E. telmateia*), 5 m (southern giant horsetail, *E. giganteum*) or 8 m (Mexican giant horsetail, *E. myriochaetum*), and allegedly even more.

Many species in this genus prefer wet sandy soils, though some are semi-aquatic and others are adapted to wet clay soils. The stalks arise from rhizomes that are deep underground and almost impossible to dig out. The field horsetail (*E. arvense*) can be a nuisance weed, readily regrowing from the rhizome after being pulled out. It is also unaffected by many herbicides designed to kill seed plants.

However, as *E. arvense* prefers an acid soil, lime may be used to assist in eradication efforts to bring the soil pH to 7 or 8. Members of the genus have been declared noxious weeds in Australia and in the US state of Oregon.

All the *Equisetum* are classed as "unwanted organisms" in New Zealand and are listed on the National Pest Plant Accord.

If eaten over a long enough period of time, some species of horsetail can be poisonous to grazing animals, including (somewhat ironically given its common name) being poisonous to horses. The toxicity appears to be related primarily to antithiaminase constituents of the plant, which lead to thiamine deficiency. People have regularly consumed horsetails. The young plants are eaten cooked or raw, but considerable care must be taken. Horsetail is dangerous for individuals with edema.

For example, the fertile stems bearing strobili of some species are cooked and eaten like asparagus (a dish called tsukushi.) in Japan. The people of ancient Rome would eat meadow horsetail in a similar manner, but they also used it to make tea as well as a thickening powder. Indians of the North American Pacific Northwest eat the young shoots of this plant raw. The plants are used as a dye and give a soft green colour. An extract is often used to provide silica for supplementation. Horsetail was often used by Indians to polish wooden tools. *Equisetum* species are often used to analyze gold concentrations in an area due to their ability to take up the metal when it is in a solution.

Anatomy

In these plants the leaves are greatly reduced and usually non-photosynthetic. They contain a single, non-branching vascular trace, which is the defining feature of microphylls. However, it has recently been recognised that horsetail microphylls are probably not primitive like in Lycopodiophyta (clubmosses and relatives), but rather advanced adaptations, evolved by the reduction of a megaphyll. They are therefore sometimes actually referred to as megaphylls to reflect this homology.

The leaves of horsetails grow in whorls fused into nodal sheaths. The stems are green and photosynthetic, and distinctive in being hollow, jointed and ridged (with sometimes 3 but usually 6-40 ridges) and these are often played with by children who will separate and then seamlessly rejoin the segments.

There may or may not be whorls of branches at the nodes; when present, these branches are identical to the main stem except being smaller and more delicate.

Spores

The spores are borne under sporangiophores in strobili, cone-like structures at the tips of some of the stems. In many species the cone-bearing stems are unbranched, and in some (e.g. field horsetail, *E. arvense*) they are non-photosynthetic, produced early in spring separately from photosynthetic sterile stems. In some other species (e.g. marsh horsetail, *E. palustre*) they are very similar to sterile stems, photosynthetic and with whorls of branches.

Horsetails are mostly homosporous, though in the field horsetail smaller spores give rise to male prothalli. The spores have four elaters that act as moisture-sensitive springs, assisting spore dispersal after the sporangia have split open longitudinally.

Systematics

Species: The living members of the genus *Equisetum* are divided into two distinct lineages, which are usually treated as subgenera. Hybrids are common, but hybridization has only been recorded between members of the same subgenus.

In addition, there are numerous ill-determined populations. One of them, the Kamchatka Horsetail (*Equisetum camtschatcense*), is an ornamental forming imposing stands of these archaic plants.

Subgenus Equisetum

- *Equisetum arvense* L. – Field Horsetail, Common Horsetail or Giant Horsetail
- *Equisetum bogotense* Kunth – Andean Horsetail or Giant Horsetail
- *Equisetum diffusum* L. – Himalayan Horsetail
- *Equisetum fluviatile* L. – Water Horsetail
- *Equisetum palustre* L. – Marsh Horsetail
- *Equisetum pratense* Ehrh. – Meadow Horsetail, Shade Horsetail, Shady Horsetail
- *Equisetum sylvaticum* L. – Wood Horsetail

Equisetum telmateia Ehrh. – Great Horsetail, Northern Giant Horsetail

Figure: *Branched Horsetail (*E. ramosissimum*)*

Subgenus *Hippochaete:*

- *Equisetum giganteum* L. – Southern Giant Horsetail or Giant Horsetail
- *Equisetum myriochaetum* Schlect. & Cham. – Mexican Giant Horsetail or Giant Horsetail
- *Equisetum hyemale* L. – Rough Horsetail, Scouringrush Horsetail
- *Equisetum laevigatum* A. Braun – Smooth Horsetail
- *Equisetum ramosissimum* Desf. – Branched Horsetail
- *Equisetum scirpoides* Michx. – Dwarf Horsetail

- *Equisetum variegatum* Schleich. ex Weber & Mohr – Variegated Horsetail

Unplaced to subgenus:

- †*Equisetum thermale* - Middle to Late Jurassic, Argentina

Named Hybrids

Figure: Equisetum × moorei *(Rough Horsetail × Branched Horsetail)*

Hybrids between species in subgenus *Equisetum*

- *Equisetum* × *bowmanii* C.N.Page (*Equisetum sylvaticum* × *Equisetum telmateia*)
- *Equisetum* × *dycei* C.N.Page (*Equisetum fluviatile* × *Equisetum palustre*)
- *Equisetum* × *font-queri* Rothm. (*Equisetum palustre* × *Equisetum telmateia*)
- *Equisetum* × *litorale* Kühlew ex Rupr. (*Equisetum arvense* × *Equisetum fluviatile*)
- *Equisetum* × *mchaffieae* C.N. Page (*Equisetum fluviatile* × *Equisetum pratense*)
- *Equisetum* × *mildeanum* Rothm. (*Equisetum pratense* × *Equisetum sylvaticum*)

- *Equisetum* × *robertsii* Dines (*Equisetum arvense* × *Equisetum telmateia*)
- *Equisetum* × *rothmaleri* C.N.Page (*Equisetum arvense* × *Equisetum palustre*)
- *Equisetum* × *willmotii* C.N.Page (*Equisetum fluviatile* × *Equisetum telmateia*)

Hybrids between Species in Subgenus Hippochaete

- *Equisetum* × *ferrissii* Clute (*Equisetum hyemale* × *Equisetum laevigatum*)
- *Equisetum* × *moorei* Newman (*Equisetum hyemale* × *Equisetum ramosissimum*)
- *Equisetum* × *nelsonii* (A.A.Eat.) Schaffn. (*Equisetum laevigatum* × *Equisetum variegatum*)
- *Equisetum* × *schaffneri* Milde (*Equisetum giganteum* × *Equisetum myriochaetum*)
- *Equisetum* × *trachydon* A.Braun (*Equisetum hyemale* × *Equisetum variegatum*)

Equisetum Cell Walls

The crude cell extracts of all *Equisetum* species tested contain mixed-linkage glucan : xyloglucan endotransglucosylase (MXE) activity. This is a novel enzyme and is not known to occur in any other plants. In addition, the cell walls of all *Equisetum* species tested contain mixed-linkage glucan (MLG), a polysaccharide which, until recently, was thought to be confined to the Poales. The evolutionary distance between *Equisetum* and the Poales suggests that each evolved MLG independently. The presence of MXE activity in *Equisetum* suggests that they have evolved MLG along with some mechanism of cell wall modification. The lack of MXE in the Poales suggests that there it must play some other, currently unknown, role. Due to the correlation between MXE activity and cell age, MXE has been proposed to promote the cessation of cell expansion.

Medicinal Uses

The plant has a long history of medicinal uses, although modern sources include cautions with regard to its use. The European Food Safety Authority issued a report assessing its medicinal uses in 2009. *Equisetum telmateia* may be a useful source of antioxidants.

Chapter 5

Poisonous Fungi

Agaricus Xanthodermus

Agaricus xanthodermus, commonly known as the yellow-staining mushroom or simply the yellow-stainer, is a mushroom of the genus *Agaricus*, which displays a strong yellow colouration at the base of the stem when cut. It is poisonous for most people, causing gastrointestinal upset, but can be eaten by some without apparent negative effect.

Taxonomy

This species was first officially defined under the name *Agaricus xanthodermus* in 1876 by Léon Gaston Genevier, in a letter published in the bulletin of the French Botanical Society. Genevier described the *Agaricus* mushrooms commonly eaten (perhaps sometimes inadvisedly) in the region of Nantes, and attempted to clarify the distinctions between them. He proposed a detailed reclassification into 5 species, including this new one. Apparently up until that time, these yellow-staining mushrooms were considered to be just varieties of other species which are edible: *A. arvensis*, *A. edulis*, and *A. silvicola*.

The epithet *xanthodermus* is derived from the Ancient Greek words for "yellow" and "skin", which were then given a Latin adjective ending. This is the official name, but the form *Agaricus xanthoderma* is also often seen. In the latter name, *-derma* "skin" is a neuter noun which does not have to agree in gender with *Agaricus*, and so this form is legal according to the rules of botanical nomenclature.

Description

The cap is generally 6–10 cm (2.4–4 in) in diameter, though can reach 15 cm (6 in). It is initially convex, with some young specimens having a squarish shape, though flattening with age. It is whitish, with light brown tints towards the centre. The cap is dry and smooth, but can be scaly when old. The gills of this mushroom progress from pale-pink to a chocolate color. The spore print is brown. Its white stipe is bulbous with a skirt-like ring. Microscopically, the cheilocystidia are club-shaped. The spores are 6-7×3-4 μm.

The main identifying feature is an immediate bright yellow colouration on cutting through the base of the stem, or scraping the flesh; later, the affected area fades to a dull brown. Numerous edible *Agaricus* species, such as *A. augustus*, *A. arvensis* and *A. silvicola*, turn yellow to a greater or lesser extent, but they do not display such an intense reaction.

Agaricus xanthodermus has an unpleasant characteristic smell, which is phenolic, reminiscent of ink or carbolic soap. The smell is especially strong at the base of the stem.

Distribution and Habitat

This mushroom is very common and widely distributed in North America, Europe, West Asia including eastern Anatolia and Iran, North Africa, and southern Africa. It has been introduced into Australia. It occurs in woods, lawns, gardens and hedgerows in autumn. This is a saprobic species.

Poisonous Character

It appears that some people can eat this mushroom without ill effects, but it can cause serious gastric problems. It is indigestible and gives rise to symptoms of sweating, flushing, and severe stomach cramps.

Of those who gather *Agaricus*-style mushrooms, about 50% of the cases of poisoning are from this species. On cooking, the smell becomes very noticeable, and this may deter people from eating it inadvertently.

Related Species

A. xanthodermus belongs to a group of related species (the "Xanthodermatei") which likewise discolour bright yellow and have a phenolic smell. They include *A. praeclaresquamosus* (formerly *A. placomyces*) which has dark grey scales, *A. moelleri*, and *A. pilatianus*, which does not have a bulbous stem.

Amanita Abrupta

Amanita abrupta, commonly known as the American abrupt-bulbed Lepidella, is a species of fungus in the Amanitaceae family of mushrooms. Named for the characteristic shape of its fruit bodies, this white *Amanita* has a slender stem, a cap covered with conical white warts, and an "abruptly enlarged" swollen base. This terrestrial species grows in mixed woods in eastern North America and eastern Asia, where it is thought to exist in a mycorrhizal relationship with a variety of both coniferous and deciduous tree species.

Taxonomy

Amanita abrupta was first described by American mycologist Charles Horton Peck in 1897, based on a specimen he found in Auburn, Alabama. Because the remains of the volva are not present on the bulb in dried, mature, specimens, Peck thought that the species should be grouped with *Amanita rubescens* and *A. spissa.* Synonyms include binomials resulting from generic transfers by Jean-Edouard Gilbert to *Lepidella* in 1928, and to *Aspidella* in 1940. Both of these genera have since been subsumed into *Amanita.*

A. abrupta is the type species of the section *Lepidella* of the genus *Amanita,* in the subgenus *Lepidella,* a grouping of related *Amanita* mushrooms characterized by their amyloid spores. Other North American species in this subgenus include *A. atkinsoniana, A. chlorinosma, A. cokeri, A. daucipes, A. mutabilis, A. onusta, A. pelioma, A. polypyramis, A. ravenelii,* and *A. rhopalopus.* European and Asian species (also in section *Lepidella*) that are phylogenetically related—close to it in the evolutionary family tree—include *A. solitaria, A. virgineoides,* and *A. japonica.*

The specific epithet *abrupta* refers to the shape of the swollen base, which is abruptly enlarged rather than gradually tapering. The species' common name is the "American abrupt-bulbed Lepidella".

Description

In *Amanita abrupta,* as with most mushrooms, the bulk of the organism lies unseen beneath the ground as an aggregation of fungal cells called hyphae; under appropriate environmental conditions, the visible reproductive structure (fruit body) is formed. The cap has a diameter of 4 to 10 centimeters (1.6 to 3.9 in), and has a broadly convex shape when young, but eventually flattens. The central portion of the cap becomes depressed in mature specimens. The cap surface is *verrucose*—covered with small angular or pyramidal erect warts (1–2 mm tall by 1–2 mm wide at the base); the warts are smaller and more numerous near the margin of the cap, and small fragments of tissue may be hanging from the margin of the cap. The cap surface, the warts, and the flesh are white. The warts can be easily separated from the cap, and in mature specimens they have often completely or partly disappeared. The white gills are placed moderately close together, reaching the stem but not directly attached to it.

The stem is 6.5 to 12.5 cm (2.6 to 4.9 in) tall, and slender, with a diameter of 0.5 to 1.5 cm (0.2 to 0.6 in). It is white, smooth (glabrous), solid (that is, not hollow internally), and has an abruptly bulbous base with the shape of a flattened sphere; it may develop longitudinal splits on the sides. The base is often attached to a copious white mycelium—a visual reminder that the bulk of the organism lies unseen below the surface. The ring is membranous, and persistent—not weathering away with time; the ring may be attached to the stem with white fibers. The mushroom has no distinct odor. The edibility of the mushroom is unknown; however, it is generally not recommended to consume *Amanita* mushrooms of questionable edibility.

Microscopic Characteristics

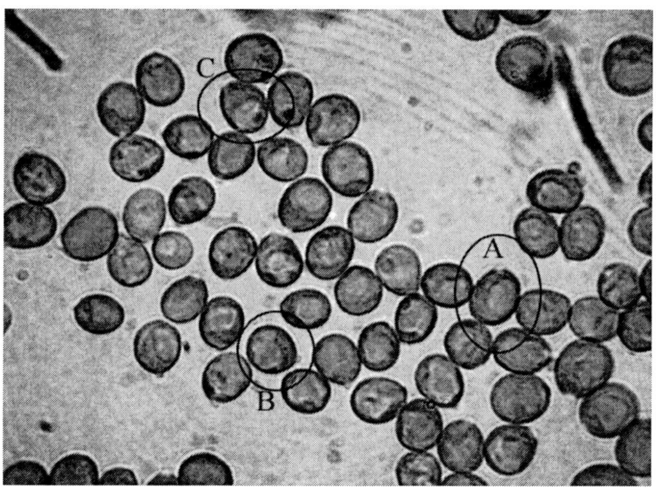

Figure: *The spores of* A. abrupta *are broadly elliptical or roughly spherical.*

When collected in deposit, such as with a spore print, the spores appear white. Viewed with a microscope, the spores are broadly elliptical or roughly spherical, smooth, thin-walled, and have dimensions of 6.5–9.5 by 5.5 by 8.5 μm. Spores are amyloid (meaning they take up iodine when stained with Melzer's reagent) The basidia (spore-bearing cells on the edges of gills) are four-spored and measure 30–50 by 4–11 μm. The bases of the basidia have clamp connections—short branches connecting one cell to the previous cell to allow passage of the products of nuclear division. The cap cuticle comprises a layer of densely interwoven, sightly gelatinized, filamentous hyphae that are 3–8 μm in diameter. The stem tissue is made of sparse, thin, longitudinally oriented hyphae measuring 294 by 39 μm.

Similar Species

The fruit bodies of *Amanita kotohiraensis*, a species known only from Japan, bears a superficial resemblance to *A. abrupta*, but *A. kotohiraensis* differs in having scattered floccose patches (tufts of soft woolly hairs that are the remains of the volva) on the cap surface, and pale yellow gills. *A. polypyramis* fruit bodies have also been noted to be similar to *A. abrupta*; however, it tends to have larger caps, up to 21 cm (8.3 in) in diameter, a fragile ring that soons withers away, and somewhat larger spores that typically measure 9–14 by 5–10 μm. The amyloidity and size of the spores are reliable characteristics to help distinguish *A. abrupta* specimens with less prominently bulbous bases from other lookalike species.

Mycologists Tsuguo Hongo and Rokuya Imazeki suggested in the 1980s that the Japanese mushroom *A. sphaerobulbosa* was synonymous with the North American *A. abrupta*. However, a 1999 study of *Amanita* specimens in Japanese herbaria concluded that they were closely related but distinct species, due to differences in spore shape and in the microstructure of the volval remnants. Another similar species, *A. magniverrucata*, is differentiated from *A. abrupta* by a number of characteristics: the universal veil is clearly separated from the flesh of the cap; the volval warts disappear more quickly because the surface of the cap cuticle gelatinizes; the partial veil is more persistent; the spores are smaller and roughly spherical; on the underside of the partial veil, the stem has surface fibrils that are drawn upward so as to somewhat resemble a cortina (a cobweb-like protective covering over the immature spore bearing surfaces); *A. magniverrucata* has a distribution limited to the eastern US.

Habitat, Distribution, and Ecology

The fruit bodies of *A. abrupta* grow on the ground, typically solitary, in mixed conifer and deciduous forests, usually during autumn. The frequency with which fruit bodies appear depends on several factors, such as season, location, temperature, and rainfall. The mushroom has been described as common in the Southeastern United States; in Texas, it has been called both infrequent, and common in the Big Thicket National Preserve. Like most other *Amanita* species, *A. abrupta* is thought to form mycorrhizal relationships with trees. This is a mutually beneficial relationship where the hyphae of the fungus grow around the roots of trees, enabling the fungus to receive moisture, protection and nutritive byproducts of the tree, and affording the tree greater access to soil nutrients. *Amanita abrupta* is widely distributed throughout eastern North America, where it has been found as far North as Quebec, Canada, and as far south as Mexico. Orson K. Miller claims to have found it in the Dominican Republic where it appeared to be growing in a mycorrizhal association with pine trees.

Kuo also mentions a mycorrhizal relationship with both hardwoods and conifers, while Tulloss lists additional preferred tree hosts such as beech, birch, fir, tsuga, oak, and poplar. However, *A. abrupta* has been shown experimentally to not form mycorrhizae with Virginia Pine.

Amanita Arocheae

Amanita arocheae, also known as the Latin American death cap is a mushroom of the large genus *Amanita*, which occurs in Central and South America. Deadly poisonous, it is a member of section phalloideae and related to the death cap *A. phalloides*. It is named after mycologist R. M. Aroche.

It is known as *hongo gris* in Mexico, where it is found under oak. It differs from the death cap in the colour of its cap, which is brownish to grayish.

Amanita Bisporigera

Amanita bisporigera is a deadly poisonous species of fungus in the Amanitaceae family. It is commonly known as the eastern North American destroying angel or the destroying angel, although it shares this latter name with three other lethal white *Amanita* species, *A. ocreata*, *A. verna* and *A. virosa*. The fruit bodies are found on the ground in mixed coniferous and deciduous forests of Eastern North America south to Mexico, but are rare in western North America; it has also been found in pine plantations in Colombia.

The mushroom has a smooth white cap that can reach up to 10 cm (3.9 in) across, and a stem, up to 14 cm (5.5 in) long by 1.8 cm (0.71 in) thick, that has a delicate white skirt-like ring near the top. The bulbous stem base is covered with a membranous sac-like volva. The white gills are free from attachment to the stalk and crowded closely together. As the species name suggests, *A. bisporigera* typically bears two spores on the basidia, although this characteristic is not as immutable as was once thought.

First described in 1906, *A. bisporigera* is classified in the section *Phalloideae* of the genus *Amanita* together with other amatoxin-containing species. Amatoxins are cyclic peptides which inhibit the enzyme RNA polymerase II and interfere with various cellular functions.

The first symptoms of poisoning appear 6 to 24 hours after consumption, followed by a period of apparent improvement, then by symptoms of liver and kidney failure, and death after four days or more. *Amanita bisporigera* closely resembles a few other white amanitas, including the equally deadly *A. virosa* and *A. verna.*

These species, difficult to distinguish from *A. bisporigera* based on visible field characteristics, do not have two-spored basidia, and do not stain yellow when a dilute solution of potassium hydroxide is applied. The DNA of *A. bisporigera* has been partially sequenced, and the genes responsible for the production of amatoxins have been determined.

Taxonomy, Classification, and Phylogeny

Amanita bisporigera was first described scientifically in 1906 by American botanist George Francis Atkinson in a publication by Cornell University colleague Charles E. Lewis. The type locality was Ithaca, New York, where several collections were made. In his 1941 monograph of world *Amanita* species, Jean-Edouard Gilbert transferred the species to his new genus *Amanitina*, but this genus is now considered synonymous with *Amanita.*

In 1944, William Murrill described the species *Amanita vernella*, collected from Gainesville, Florida; that species is now thought to be synonymous with *A. bisporigera* after a 1979 examination of its type material revealed basidia that were mostly 2-spored. *Amanita phalloides* var. *striatula*, a poorly known taxon originally described from the United States in 1902 by Charles Horton Peck, is considered by *Amanita* authority Rodham Tulloss to be synonymous with *A. bisporigera.* Vernacular names for the mushroom include "destroying angel", "deadly amanita", "white death cap", "angel of death" and "eastern North American destroying angel".

Amanita bisporigera belongs to section *Phalloideae* of the genus *Amanita*, which contains some of the deadliest *Amanita* species, including *A. phalloides* and *A. virosa.* This classification has been upheld with phylogenetic analyses, which demonstrate that the toxin-producing members of section *Phalloidae* form a clade—that is, they derive from a common ancestor. In 2005, Zhang and colleagues performed a phylogenetic analysis based on the internal transcribed spacer (ITS) sequences of several white-bodied toxic *Amanita* species, most of which are found in Asia. Their results support a clade containing *A. bisporigera*, *A. subjunquillea* var. *alba*, *A. exitialis*, and *A. virosa*. The Guangzhou destroying angel (*Amanita exitialis*) has two-spored basidia, like *A. bisporigera*.

Description

The cap is 3–10 cm (1.2–3.9 in) in diameter and, depending on its age, ranges in shape from egg-shaped to convex to somewhat flattened. The cap surface is smooth and white, sometimes with a pale tan- or cream-colored tint in the center. The surface is either dry or, when the environment is moist, slightly sticky. The flesh is thin and white, and does not change color when bruised.

The margin of the cap, which is rolled inwards in young specimens, does not have striations (grooves), and lacks volval remnants. The gills, also white, are crowded closely together. They are either free from attachment to the stem or just barely reach it. The lamellulae (short gills that do not extend all the way to the stem) are numerous, and gradually narrow.

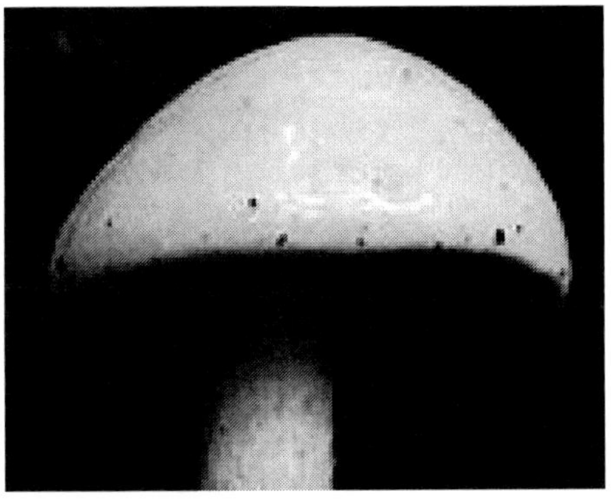

Figure: *Surfaces turn yellow when stained with potassium hydroxide solution.*

The white stem is 6–14 cm (2.4–5.5 in) by 0.7–1.8 cm (0.3–0.7 in) thick, solid (i.e., not hollow), and tapers slightly upward. The surface, in young specimens especially, is frequently floccose (covered with tufts of soft hair), fibrillose (covered with small slender fibers), or squamulose (covered with small scales); there may be fine grooves along its length. The bulb at the base of the stem is spherical or nearly so. The delicate ring on the upper part of the stem is a remnant of the partial veil that extends from the cap margin to the stalk and covers the gills during development. It is white, thin, membranous, and hangs like a skirt. When young, the mushrooms are enveloped in a membrane called the universal veil, which stretches from the top of the cap to the bottom of the stem, imparting an oval, egg-like appearance. In mature fruit bodies, the veil's remnants form a membrane around the base, the volva, like an eggshell-shaped cup.

On occasion, however, the volva remains underground or gets torn up during development. It is white, sometimes lobed, and may become pressed closely to the stem. The volva is up to 3.8 cm (1.5 in) in height (measured from the base of the bulb), and is about 2 mm thick midway between the top and the base attachment. The mushroom's odor has been described as "pleasant to somewhat nauseous", becoming more cloying as the fruit body ages.

The cap flesh turns yellow when a solution of potassium hydroxide (KOH, 5–10%) is applied (a common chemical test used in mushroom identification). This characteristic chemical reaction is shared with *A. ocreata* and *A. virosa*, although some authors have expressed doubt about the identity of North American *A. virosa*, suggesting those collections may represent four-spored *A. bisporigera*. Tulloss suggests that reports of *A. bisporigera* that do not turn yellow with KOH were actually based on white forms of *A. phalloides*. Findings from the Chiricahua Mountains of Arizona and in central Mexico, although "nearly identical" to *A. bisporigera*, do not stain yellow with KOH; their taxonomic status has not been investigated in detail.

Microscopic Features

The spore print of *A. bisporigera*, like most *Amanita*, is white. The spores are roughly spherical, thin-walled, hyaline (translucent), amyloid, and measure 7.8–9.6 by 7.0–9.0 ìm. The cap cuticle is made of partially gelatinized, filamentous interwoven hyphae, 2–6 ìm in diameter. The tissue of the gill is bilateral, meaning it diverges from the center of the gill to its outer edge. The subhymenium is ramose—composed of relatively thin branching, unclamped hyphae.

The spore-bearing cells, the basidia, are club-shaped, thin-walled, without clamps, with dimensions of 34–45 by 4–11 ìm. They are typically two-spored, although rarely three- or four-spored forms have been found. Although the two-spored basidia are a defining characteristic of the species, there is evidence of a tendency to shift towards producing four-spored basidia as the fruiting season progresses.

The volva is composed almost exclusively of densely interwoven filamentous hyphae, 2–10 ìm in diameter, that are sparsely to moderately branched. There are few small inflated cells, which are mostly spherical to broadly elliptic. The tissue of the stem is made of abundant, sparsely branched, filamentous hyphae, without clamps, measuring 2–5 ìm in diameter.

The inflated cells are club-shaped, longitudinally oriented, and up to 2–3 by 15.7 ìm. The annulus is made of abundant moderately branched filamentous hyphae, measuring 2–6 ìm in diameter. The inflated cells are sparse, broadly elliptic to pear-shaped, and are rarely larger than 31 by 22 ìm. Pleurocystidia and cheilocystidia (cystidia found on the gill faces and edges, respectively) are absent, but there may be cylindrical to sac-like cells of the partial veil on the gill edges; these cells are hyaline and measure 24–34 by 7–16 ìm.In 1906 Charles E. Lewis studied and illustrated the development of the basidia in order to compare the nuclear behavior of the two-spored with that of the four-spored forms.

Initially (1), the young basidium, appearing as a club-shaped branch from the subhymenium, is filled with cytoplasm and contains two primary nuclei, which have distinct nucleoli. As the basidium grows larger, the membranes of the two nuclei contact (2), and then the membrane disappears at the point of contact (3). The two primary nuclei remain distinct for a short time, but eventually the two nuclei fuse completely to form a larger secondary nucleus with a single secondary nucleolus (4, 5). The basidium increases in size after the primary nuclei fuse, and the nucleus migrates towards the end of the basidia (6, 7). During this time, the nucleus develops vacuoles "filled by the nuclear sap in the living cell". Chromosomes are produced from the nucleolar threads, and align transversely near the apex of the basidium, connected by spindles (8–10). The chromosomes then move to the poles, forming the daughter nuclei that occupy different positions in the basidium; the daughters now have a structure similar to that of the parent nuclei (11). The two nuclei then divide to form four nuclei, similar to fungi with four-spored basidia (12, 13). The four nuclei crowd together at some distance from the end of the basidium to form an irregular mass (14). Shortly thereafter, the sterigmata (slender projections of the basidia that attach the spores) begin to form (15), and cytoplasm begins to pass through the sterigmata to form the spores (16). Although Lewis was not able to clearly determine from observation alone whether the contents of two or four nuclei passed through the sterigmata, he deduced, by examining older basidia with mature spores, that only two nuclei enter the spores (16, 17).

Toxicity

Amanita bisporigera is considered the most toxic North American *Amanita* mushroom, with little variation in toxin content between different fruit bodies. Three subtypes of amatoxin have been described: á-, â, and ã-amanitin. The principal amatoxin, á-amanitin, is readily absorbed across the intestine, and 60% of the absorbed toxin is excreted into bile and undergoes enterohepatic circulation; the kidneys clear the remaining 40%. The toxin inhibits the enzyme RNA polymerase II, thereby interfering with DNA transcription, which suppresses RNA production and protein synthesis.

This causes cellular necrosis, especially in cells which are initially exposed and have rapid rates of protein synthesis. This process results in severe acute liver dysfunction and, ultimately, liver failure. Amatoxins are not broken down by boiling, freezing, or drying. Roughly 0.2 to 0.4 milligrams of á-amanitin is present in 1 gram of *A. bisporigera*; the lethal dose in humans is less than 0.1 mg/kg body weight. One mature fruit body can contain 10–12 mg of á-amanitin, enough for a lethal dose.

The á-amanitin concentration in the spores is about 17% that of the fruit body tissues. *A. bisporigera* also contains the phallotoxin phallacidin, structurally related to the amatoxins but considered less poisonous because of poor absorption. Poisonings (from similar white amanitas) have also been reported in domestic animals, including dogs, cats, and cows.

The first reported poisonings resulting in death from the consumption of *A. bisporigera* were from near San Antonio, Mexico in 1957, where a rancher, his wife, and three children consumed the fungus; only the man survived. *Amanita* poisoning is characterized by the following distinct stages: The *incubation stage* is an asymptomatic period which ranges from 6 to 12 hours after ingestion.

In the *gastrointestinal stage*, about 6 to 16 hours after ingestion, there is onset of abdominal pain, explosive vomiting, and diarrhea for up to 24 hours, which may lead to dehydration, severe electrolyte imbalances, and shock. These early symptoms may be related to other toxins such as phalloidin. In the cytotoxic stage, 24 to 48 hours after ingestion, clinical and biochemical signs of liver damage are observed, but the patient is typically free of gastrointestinal symptoms. The signs of liver dysfunction such as jaundice, hypoglycemia, acidosis, and hemorrhage appear. Later, there is an increase in the levels of prothrombin and blood levels of ammonia, and the signs of hepatic encephalopathy and/or kidney failure appear. The risk factors for mortality that have been reported are age younger than 10 years, short latency period between ingestion and onset of symptoms, severe coagulopathy (blood clotting disorder), severe hyperbilirubinemia (jaundice), and rising serum creatinine levels.

Similar Species

The color and general appearance of *A. bisporigera* are similar to those of *A. verna* and *A. virosa*. *A. bisporigera* is at times smaller and more slender than either *A. verna* or *A. virosa*, but it varies considerably in size; therefore size is not a reliable diagnostic characteristic. *A. virosa* fruits in autumn—later than *A. bisporigera*. *A. elliptosperma* is less common but widely distributed in the southeastern United States, while *A. ocreata* is found on the West Coast and in the Southwest. Other similar toxic North American species include *Amanita magnivelaris*, which has a cream-colored, rather thick, felted-submembranous, skirt-like ring, and *A. virosiformis*, which has elongated spores that are 3.9–4.7 by 11.7–13.4 ìm. Neither *A. elliptosperma* nor *A. magnivelaris* typically turn yellow with the application of KOH; the KOH reaction of *A. virosiformis* has not been reported.

Leucoagaricus leucothites is another all-white mushroom with an annulus, free gills, and white spore print, but it lacks a volva and has thick-walled dextrinoid (staining red-brown in Melzer's reagent) egg-shaped spores with a pore. *A. bisporigera* may also be confused with the larger edible species *Agaricus silvicola*, the "horse-mushroom". Like many white amanitas, young fruit bodies of *A. bisporigera*, still enveloped in the universal veil, can be confused with puffball species, but a longitudinal cut of the fruit body reveals internal structures in the *Amanita* that are absent in puffballs. In 2006, seven members of the Hmong community living in Minnesota were poisoned with *A. bisporigera* because they had confused it with edible paddy straw mushrooms (*Volvariella volvacea*) that grow in Southeast Asia.

Habitat and Distribution

Like most other *Amanita* species, *A. bisporigera* is thought to form mycorrhizal relationships with trees. This is a mutually beneficial relationship where the hyphae of the fungus grow around the roots of trees, enabling the fungus to receive moisture, protection and nutritive byproducts of the tree, and giving the tree greater access to soil nutrients. Fruit bodies of *Amanita bisporigera* are found on the ground growing either solitarily, scattered, or in groups in mixed coniferous and deciduous forests; they tend to appear during summer and early fall. The fruit bodies are commonly found near oak, but have been reported in birch-aspen areas in the west.

It is most commonly found in eastern North America, and rare in western North America. It is widely distributed in Canada, and its range extends south to Mexico. The species has also been found in Colombia, where it may have been introduced from trees exported for use in pine plantations.

Genome Sequencing

The *Amanita* Genome Project was begun in Jonathan Walton's lab at Michigan State University in 2004 as part of their ongoing studies of *Amanita bisporigera*. The purpose of the project is to determine the genes and genetic controls associated with the formation of mycorrhizae, and to elucidate the biochemical mechanisms of toxin production. The genome of *A. bisporigera* has been sequenced using a combination of automated Sanger sequencing and pyrosequencing, and the genome sequence information is publicly searchable.

The sequence data enabled the researchers to identify the genes responsible for amatoxin and phallotoxin biosynthesis, *AMA1* and *PHA1*. The cyclic peptides are synthesized on ribosomes, and require proline-specific peptidases from the prolyl oligopeptidase family for processing.

The genetic sequence information from *A. bisporigera* has been used to identify molecular polymorphisms in the related *A. phalloides*. These single-nucleotide polymorphisms may be used as population genetic markers to study phylogeography and population genetics. Sequence information has also been employed to show that *A. bisporigera* lacks many of the major classes of secreted enzymes that break down the complex polysaccharides of plant cell walls, like cellulose.

In contrast, saprobic fungi like *Coprinopsis cinerea* and *Galerina marginata*, which break down organic matter to obtain nutrients, have a more complete complement of cell wall-degrading enzymes. Although few ectomycorrhizal fungi have yet been tested in this way, the authors suggest that the absence of plant cell wall-degrading ability may correlate with the ectomycorrhizal ecological niche.

Amanita Brunnescens

Amanita brunnescens, also known as the Brown American Star-Footed Amanita or cleft-footed amanita is a native North American mushroom of the large genus *Amanita*. Originally presumed to be *Amanita phalloides* by renowned American mycologist Charles Horton Peck it was described and named by G. F. Atkinson of Cornell University. He named it after the fact that it bruised brown.

It differs from the death cap by its fragile volva and tendency to bruise brown.

Amanita Exitialis

Amanita exitialis, also known as the Guangzhou destroying angel, is a mushroom of the large genus *Amanita*. It is distributed in eastern Asia, and probably also in India where it has been misidentified as *A. verna*. Deadly poisonous, it is a member of section Phalloideae and related to the death cap *A. phalloides*.

The fruit bodies (mushrooms) are white, small to medium-sized with caps up to 7 cm (2.8 in) in diameter, a somewhat friable ring and a firm volva. Unlike most agaric mushrooms which typically have four-spored basidia (spore-bearing cells), the basidia of *A. exitialis* are almost entirely two-spored. Eight people were fatally poisoned in China after consuming the mushroom in 2000, and another 20 have been fatally poisoned since that incident. Molecular analysis shows that the species has a close phylogenetic relationship with three other toxic white Amanitas: *A. subjunquillea* var. *alba*, *A. virosa* and *A. bisporigera*.

Taxonomy, Classification, and Phylogeny

Zhu-Liang Yang and Tai-Hui Li discovered the species by reexamining various herbarium specimens of white *Amanita* typically referred to as either *A. verna* or *A. virosa*. They realized that collections referred to as these European species actually comprised three taxa new to science or the region. The holotype specimen of *A. exitialis* is located in the Mycological Herbarium of Guangdong Institute of Microbiology.

In 2005, Zhang and colleagues performed a phylogenetic analysis based on the ITS sequences of several white-bodied toxic *Amanita* species. Their results support a clade containing four lethal *Amanita* species with white fruit bodies. *A. exitialis* has two-spored basidia similar to the North American species *A. bisporigera*, but *A. exitialis* has a closer phylogenetic relationship with *Amanita subjunquillea* var.

alba, a four-spored white lethal species from China. The specific epithet *exitialis* derives from the Latin word "destructive", and refers to the lethally poisonous nature of the mushroom.

Amanita exitialis is classified in the section Phalloideae of the genus *Amanita*. Species of this section share the following characteristics: spores that are not strongly elongated, and never with a cyclindric shape; flesh not reddening upon bruising; lamellulae (short gills that do not reach the edge of the cap) abruptly cut off; a well-formed pouch- or sac-like membranous volva.

All species of Phalloideae are ectomycorrhizal and contain amatoxins.

Description

The cap is 4–7 cm (1.6–2.8 in) in diameter, initially egg-shaped, then convex but flattening with age, and sometimes slightly depressed at the center. The cap surface is smooth, white, but cream-colored in the center. The margin (cap edge) is non-striate, non-appendiculate (without any partial veil remnants hanging along the cap margin); the flesh white. The gills are free from attachment to the stem, white to whitish, crowded closely together, and up to 5 mm in height.

The lamellulae are long and tapering, plentiful, and arranged in 2–3 tiers. The stem is 7–9 cm (2.8–3.5 in) by 0.5–1 cm (0.2–0.4 in), roughly cylindrical or slightly tapering upward, with apex slightly expanded. The surface is white to whitish, smooth, or sometimes with fibrous small scales. The bulb at the base of the stem is roughly spherical and 1–2 cm (0.4–0.8 in) wide.

The volva is limbate (has a distinct edge), thin, membranous, with free limb up to 7 mm in height, and both surfaces are white. The ring is near the top of the stem, thin, membranous, white, persistent or may be torn from the stem during expansion of the cap. All tissues of the fruit body will turn yellow if a drop of dilute potassium hydroxide is applied.

The spores are spherical or nearly so, rarely broadly ellipsoid, and measure 9.5–12 by 9–11.5 μm. They are hyaline (translucent) and colorless, amyloid (absorbing iodine when stained with Melzer's reagent), thin-walled, smooth, and have a small apiculus.

The spore-bearing cells, the basidia, are 27–55 by 10–15 μm, club-shaped, and two-spored (rarely one-spored). They have sterigmata (slender projections that attach the spores) that are 5–7 μm long.

Similar Species

Amanita exitialis is similar to *A. bisporigera*, a species originally described by George Francis Atkinson from the United States. In comparison to *A. exitialis*, *A. bisporigera* differs by its lower placement of the ring on the stem, smaller spores (typically 8–9.5 by 7–8.5 μm), and different structure of the volva.

American specimens of *A. bisporigera* have more abundant inflated cells than that of *A. exitialis*. Two other white Asian species, *A. oberwinklerana* and *A. subjunquillea* var. *alba* also resemble *A. exitialis*, but are four-spored.

Habitat and Distribution

The fruit bodies of *A. exitialis* grow solitarily or in groups on the ground in coniferous forests. It is known only from the type locality, Guangdong Province. In a 2003 survey of Indian *Amanita* species, the authors mention several collections identified as *A. verna*, from various locations in the Indian provinces Maharashtra, Orissa, and Rajasthan. As Bhatt *et al.* explain, the material collected by Dhanchiola in Orissa and identified as *A. virosa* has two-spored basidia, and his description matches that of *A. exitialis*.

Field observations suggest that the mushroom associates mycorrhizally with the plant *Castanopsis fissa*, a deciduous tree found only in the southern provinces of China, such as Guangdong, Yunnan and Hunan. Mushroom fruitings are abundant in the warm spring rains of March and April, although they are also seen in May to July.

Toxicity

The content and distribution of the main amatoxins (alpha-amanitin, beta-amanitin) and phallotoxins (phallacidin, phallisin, phalloin, phalloidin) in the three tissues (cap, stipe and volva) of *Amanita exitialis* have been determined using high-performance liquid chromatography.

The cap had the highest content of total toxins, reaching over 8000 μg/g dry weight (μg of toxin per gram of dried tissue), the toxins content in the stem was over 3700 μg/g dry weight, whereas the volva had the lowest content of total toxins, with about 1150 μg/g dry weight. Amatoxins content (alpha-amanitin and beta-amanitins especially alpha-amanitin) in the cap, stem or volva of *A. exitialis* was higher than that of Phallotoxins (Phallacidin, Phallisin, Phalloidin and Phalloin), but the content of phallotoxins (especially phallacidin) was gradually higher from cap to stem and to volva. A 2011 study reported the presence of additional toxins amaninamide, phallacin, phallisacin and desoxoviridin.

A Chinese study concluded that this species had the highest mortality rate of all toxic mushrooms in China. It has been estimated that about 50 grams (1.8 oz) of fresh mushrooms contains sufficient toxin to cause the death of a 50-kilogram (110 lb) adult. In March 2000, nine people consumed the mushroom in Guangzhou, and only one survived. Since 2000, another 20 people have died in the southern provinces of China from consuming the mushroom.

Toxic peptides from *Amanita* species have been widely used in biological research as chemical agents to inhibit RNA polymerase II, an enzyme essential for protein synthesis. However, these toxic peptides can only be obtained from fruit bodies collected from natural habitats, and are consequently expensive. Some success has been reported in extracting peptide toxins directly from the mycelia of *Amanita exitialis* grown in liquid culture.

Although the toxin concentration in the mycelium is only about 10% of that in fruiting bodies, the authors suggest that is possible to increase the amatoxin production by optimizing the growth conditions.

Bioactive Compounds

The fruit bodies of *Amanita exitialis* contain a unique purine nucleoside that is coupled with an amino acid derivative named N^2-(1-methoxycarbonylethyl)guanosine. The discovery and identification of this chemical was the first report of a naturally occurring nucleoside in which an amino acid derivative is bonded through its á-amino nitrogen (the nitrogen bonded to the á-carbon) to a nucleobase aglycone by a C-N (carbon to nitrogen) bond.

The new compound was determined to be toxic in the brine shrimp lethality test, but it did not have cytotoxic activity against a variety of human cancer cell lines. Other compounds isolated from the fungus include â-carboline and russulaceramide (a ceramide formerly found in some *Russula* mushrooms).

Amanita Farinosa

Amanita farinosa, also called Eastern American Floury Amanita, is a North American poisonous mushroom of the genus *Amanita*, a genus of fungi including some of the most deadly mushrooms, as well as notably psychoactive mushrooms.

Taxonomy

Two recent molecular studies show that *Amanita farinosa* is part of a subgroup within *Amanita* with its close relatives *Amanita muscaria, A. gemmata* and *A. roseotincta.*

Physical Description

- *Cap :* The cap is 2.5–7 cm (1-2.8 inches) in diameter, domed in young and flat in older specimens, with a striate margin. It is whitish grey and covered with brownish grey volval or mealy material.
- *Gills :* The gills are white. They may be attached to the stem or free. They are close and crowded and not bruising.
- *Stem / Stipe :* The stem, or the stipe, lacks a ring and at its base a smallish bulb or volva. It measures up to 6.5 cm high, 1-3 centimeters thick. The stem is white to tan in color.
- *Spores and microscopic features :* The spores are white. They are 5.5-8 x 6-8 μ in measurement and inamyloid. The spores are smooth and round to broadly elliptical.
- *Flesh :* The flesh is white in color. It doesn't stain on exposure.
- *Odor :* In old specimens, the smell can be strong and that of mink.

Distribution and Habitat

An uncommon mushroom, it is found across North America in late summer to late autumn in coniferous or deciduous woodlands.

Amanita Gemmata

Amanita gemmata, commonly known as the gemmed Amanita, jeweled deathcap or the jonquil Amanita, is an agaric mushroom of the family Amanitaceae and genus *Amanita.* The fruit body has a cap that is a dull to golden shade of yellow, and typically 2.5–12 cm (1.0–4.7 in) in diameter. The cap surface is sticky when moist, and characterized by white warts, which are easily detached. It is initially convex, and flattens out when mature.

The flesh is white and does not change colour when cut. Gills are white and closely spaced. The stem is pale yellow, and measures 4–12 cm (1.6–4.7 in) long by 0.5–1.9 cm (0.2–0.7 in) thick. The partial veil that covers the young fruit body turns into the ring on the stem at maturity. The spore print is white, while the spores are roughly elliptical, and measure 8–10 by 6.5–7.5 μm.

This species is a mycorrhizal fungus, widespread in the Americas and Europe. It can grow either singly, scattered, or in groups. It prefers habitats like coniferous and mixed forests and alongside paths, where it fruits in summer and fall.

Figure: *Amanita Gemmata*

It is a toxic mushroom that results in various symptoms of poisoning after consumption, but deaths are rare. It is often confused with various other European species.

A. gemmata resembles the false death cap, tawny grisette and panther cap mushrooms. Its cap is brighter in color than the former, and more yellow than the latter two.

Taxonomy and Phylogeny

he species was first described scientifically by Swedish mycologist and botanist Elias Magnus Fries as *Agaricus gemmatus* in 1838. It was transferred to the genus *Amanita* in 1866 by the French statistician Louis Bertillon.

The species has been transferred to several genera in its history, resulting in a number of synonyms, including *Amanita muscaria* var. *gemmata* (1886, Lucien Quélet), *Amanitopsis gemmata* (1887, Pier Andrea Saccardo), *Amanitaria gemmata* (1940, Jean-Edouard Gilbert), and *Venenarius gemmatus* (1948, William Murrill).

Amanita authority Rodham E. Tulloss considers *A. amici* (published by Claude Casimir Gillet in 1891) to be synonymous with *A. gemmata*, as the macroscopic characteristics of the former fall within the limits of the range expected for the latter.

Within the genus *Amanita*, *A. gemmata* is classified in subgenus *Amanita*, section *Amanita*, subsection *Gemmatae*, and series *Gemmatae*. Tulloss places the species in a stirps (an informal ranking below species level) with *A. russuloides* and *A. viscidolutea*.

Some mycologists believe that *A. gemmata* is not different from *A. russuloides*. Two molecular studies based on sequences of the large ribosomal subunit RNA gene (nLSU-rDNA) and the mitochondrial small ribosomal subunit RNA gene (mtSSU-rDNA) show that *A. gemmata* is part of a clade within *Amanita* with its close relatives *A. muscaria*, *A. farinosa* and *A. roseitincta*.

The mushroom is commonly known as the "gemmed Amanita", the "jonquil Amanita", the "jeweled death cap", or the "European gemmed Amanita".

Description

Figure: *Young specimen after several days of dry weather (note pale color of dry cap)*

The fruit bodies are colored yellow overall. The fresh cap, ranging in color from dull creamy yellow to golden yellow to buff, is sticky when moist. White warts adorn the cap surface, but they are usually flimsy and easily washed away by rain. They are placed randomly, but tend to be more concentrated in the center. The cap is typically 2.5–12 cm (1.0–4.7 in) in diameter, and initially convex before flattening out in maturity. The flesh is white, and shows no change when sliced.

The gills are adnate to adnexed, and white; they are close together, with little intervening space. The pale yellowish stem is 4–12 cm (1.6–4.7 in) long by 0.5–1.9 cm (0.2–0.7 in) thick, and either roughly equal in width throughout, or slightly thicker at the base. Young mushrooms have a membranous partial veil extending from the upper stem to the cap margin; as the mushroom grows, the partial veil tears to leave a flimsy, skirt-like, easily-lost ring on the stem. At the base of the stem is a white volva (a remnant of the universal veil that covered the immature mushroom) that usually forms a small, free rim. Spore prints are white. There is no distinctive odour.

Microscopic Characteristics

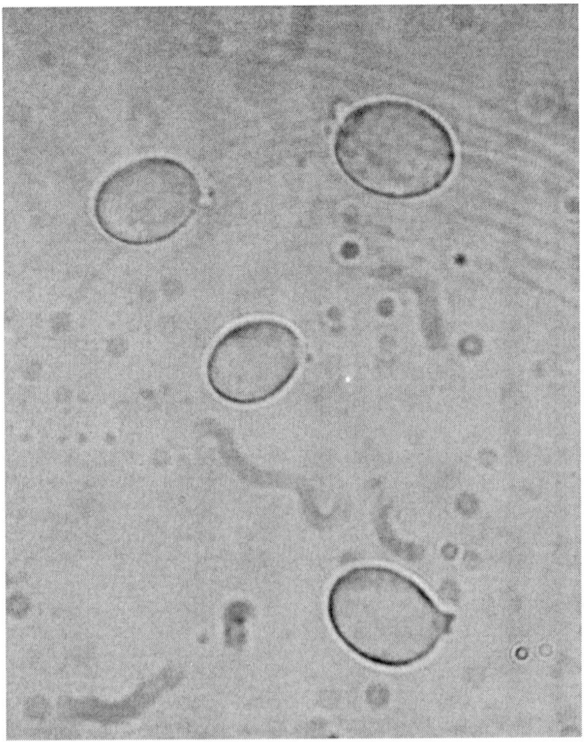

Figure: *The spores are thin-walled and ellipsoid.*

Amanita gemmata has ellipsoid to broadly ellipsoid spores measuring 8–10 by 6.5–7.5 µm with an average Q-ratio (the fraction of length/width) of 1.35; they are not amyloid. The spores are smooth, thin-walled, and they contain one to several small oil droplets. The basidia (spore-bearing cells of the hymenium) are usually four-spored, club shaped, and measure 30–40 by 8–11 µm.

The gill tissue is *divergent*, meaning that the cells are more or less parallel near the center of the gill, but bend outwards near the end of the gill. The hyphae in this tissue are cylindrical to inflated, thin walled, hyaline (translucent) to yellowish, and measure 2.2–9 µm wide; the hyphae in the central strand are narrower and typically cylindrical. The hyphae of the subhymenium (a layer of tissue directly under the hymenium) are interwoven. These hyphae are branched, cylindrical to slightly inflated, hyaline, and 6–9 µm wide.

The hyphae of the cap cuticle are filamentous, interwoven, and radially arranged. They are cylindrical, 2.7–4 µm wide, thin-walled, hyaline to yellowish, and gelatinize when mounted in potassium hydroxide. The cap tissue is also interwoven, with hyphae that are cylindrical to somewhat inflated, 3.7–14.6 µm wide, thin-walled, branched, and hyaline to yellowish. Caulocystidia are abundant on the apex of the stem; they are club-shaped to cylindrical, thin-walled, hyaline, and measure 3–9 µm wide.

The annulus tissue comprises interwoven cylindrical hyphae measuring 3–9 µm wide. Sphaerocysts (inflated, spherical cells) are also present in the annulus tissue; they are club shaped to ellipsoidal, with dimensions of 29–55 by 30–70 µm. The warts on the cap surface (remnants of the universal veil) comprise loosely interwoven cylindrical to inflated thin-walled hyphae that are 3.5–8 µm wide. Sphaerocysts in this tissue are 58.5–70.2 by 17.5–40 µm, ellipsoidal, and hyaline. The volval tissue is interwoven, with cylindrical, hyaline hyphae that are 4.4–7.3 µm wide. The sphaerocysts here are ellipsoidal to roughly spherical, hyaline, and measure 35–70 by 20–35 µm. In *A. gemmata*, where they are most abundant in the region just below the cap cuticle, these refractive cells are scattered, and have a width of 3.7–6 µm. Clamp connections are rare in the hyphae of *A. gemmata*; they are present in the annulus, gill tissue, subhymenium, and cap tissue.

Similar Species

There are numerous forms in North America that tend to intergrade with *A. pantherina*. In 2005, mycologist Rod Tulloss described *Amanita aprica*, a species that has been confused with *A. gemmata* several times in the past. According to mycologists Pierre Neville and Serge Poumarat, the Mediterranean species *A. amici* (synonymous with *A. gemmata* f. *amici*) is similar in appearance to *A. gemmata* but is larger. According to Tulloss however, their measurements of the cap and stem dimensions of *A. amici* fell within the range expected for *A. gemmata*, and for this reason, the two taxa should be considered conspecific. Neville and Poumarat suggest that the name *A. gemmata* still persists for Mediterranean collections because of its frequent historical misapplication to the native Italian species *A. gioiosa*, which had not been described as a distinct species until 2004. *A. orientigemmata*, a mushroom ranging from Japan to China, is a lookalike, but has clamps, unlike *A. gemmata*. Other differences between the two species include the slightly smaller spores of *A. orientigemmata*, and differences in the microstructure of the cap warts.

Toxicity

The toxicity of *Amanita gemmata* may be due to its possible hybridization with *A. pantherina*, which produces forms with characteristics intermediate between the two. Toxicity is suspected to be due to the presence of muscimol and ibonetic acid. So it also causes pantherina syndrome. The species was implicated in several poisonings (including one death) in the Chilean province of Malleco between 1986 and 1990. Symptoms were similar to those caused by ingestion of *Amanita phalloides*, including acute gastroenteritis and acute hepatitis.

Generally, symptoms of poisoning appear within three hours of ingestion of the mushroom as visual hallucinations, nausea, vomiting, stomach pain, diarrhea, irregular and slow heart beat and agitation. Severe cases involving coma, convulsions, or death are extremely rare.

Ecology, Habitat and Distribution

Amanita gemmata is a mycorrhizal fungus, meaning it forms a mutually beneficial relationship with the roots of compatible host plants. Through the association, the plant provides the fungus with a carbon source, and the fungus provides the plant with several benefits such as nutrients and protection from pathogens. Largent and collaborators (1980) document mycorrhizal associations of *A. gemmata* with Manzanita (*Arctostaphylos* spp.) and Lodgepole Pine (*Pinus contorta*), and Nieto and Carbone with Maritime Pine (*Pinus pinaster*) in Spain. The fungus favours sandy and slightly acidic soils, and is often found in association with Norway Spruce (*Picea abies*). The mushroom grows either singly, scattered, or in groups in coniferous and mixed forests, especially along paths and roads.

Figure: *A. gemmata specimen from Commanster, Belgium*

The species is distributed in areas of Asia, the Americas, and Europe. It fruits in summer and fall (fall and winter in California). It is widely distributed (as a species cluster) in North America, where it has been found as far south as Ixtlán de Juárez, Mexico. The species has been reported from the Dominican Republic. In South America, it is known from Chile. In Asia, the mushroom has been collected from Iran and China.

Amanita Elliptosperma

Amanita elliptosperma, commonly known as the Atkinson's destroying angel, is a basidiomycete fungus, one of many in the genus *Amanita*. Although its toxicity is unknown, it is likely to be deadly poisonous like its close relatives. Originally described from North Carolina, it's found down the eastern United States from New England to eastern Texas.

Amanita Muscaria

Amanita muscaria, commonly known as the fly agaric (pron.: / ÈæaYrjk/) or fly amanita (pron.: /ÌæmYÈnajtY/), is a poisonous and psychoactive basidiomycete fungus, one of many in the genus *Amanita*. Native throughout the temperate and boreal regions of the Northern Hemisphere, *Amanita muscaria* has been unintentionally introduced to many countries in the southern hemisphere, generally as a symbiont with pine plantations, and is now a true cosmopolitan species. It associates with various deciduous and coniferous trees. The quintessential toadstool, it is a large white-gilled, white-spotted, usually deep red mushroom, one of the most recognizable and widely encountered in popular culture. Several subspecies with differing cap colour have been recognised to date, including the brown *regalis* (considered a separate species), the yellow-orange *flavivolvata, guessowii, formosa,* and the pinkish *persicina.* Genetic studies published in 2006 and 2008 show several sharply delineated clades that may represent separate species.

Figure: *Amanita Muscaria: Showing three stages as the mushroom expands*

Although it is generally considered poisonous, deaths from its consumption are extremely rare, and it is eaten as a food in parts of Europe, Asia, and North America after parboiling. *Amanita muscaria* is noted for its hallucinogenic properties, with its main psychoactive constituent being the compound muscimol. It was used as an intoxicant and entheogen by the peoples of Siberia and has a religious significance in these cultures. There has been much speculation on traditional use of this mushroom as an intoxicant in places other than Siberia; however, such traditions are far less well documented. The American banker and amateur ethnomycologist R. Gordon Wasson proposed that the fly agaric was in fact the soma of the ancient Rig Veda texts of India; since its introduction in 1968 this theory has gained both followers and detractors in anthropological literature.

Taxonomy and Naming

The name of the mushroom in many European languages is thought to be derived from its use as an insecticide, when sprinkled in milk.

This practice has been recorded from Germanic- and Slavic-speaking parts of Europe, as well as the Vosges region and pockets elsewhere in France, and Romania. Albertus Magnus was the first to record it in his work *De vegetabilibus* sometime before 1256, commenting *vocatur fungus muscarum, eo quod in lacte pulverizatus interficit muscas*, "it is called the fly mushroom because it is powdered in milk to kill flies."

Figure: *Showing the partial veil under the cap dropping away to form a ring around the stipe*

The 16th century Flemish botanist Carolus Clusius traced the practice of sprinkling it into milk to Frankfurt in Germany, while Carl Linnaeus, the "father of taxonomy", reported it from Småland in southern Sweden where he had lived as a child. He officially described it in Volume Two of his *Species Plantarum* in 1753, giving it the name *Agaricus muscarius*, the specific epithet deriving from Latin *musca* meaning "fly". It gained its current name in 1783, when placed in the genus *Amanita* by Jean-Baptiste Lamarck and sanctioned by Elias Magnus Fries.

The starting date was set as January 1, 1821, the date of the works of the "father of mycology", Swedish naturalist Elias Magnus Fries, and so, the full name was *Amanita muscaria* (L.:Fr.) Hook.. However, the 1987 edition of the International Code of Botanical Nomenclature changed the rules regarding the starting date and primary work for names of fungi, and now names can be considered valid as far back as May 1, 1753, the date of publication of Linnaeus's seminal work. Hence, Linnaeus and Lamarck became the namers of the *Amanita muscaria* (L.) Lam.

The English mycologist John Ramsbottom reported that *Amanita muscaria* was used for getting rid of bugs in England and Sweden, and *bug agaric* was an old alternate name. French mycologist Pierre Bulliard reported having tried without success to replicate its fly-killing properties in his work *Histoire des plantes vénéneuses et suspectes de la France*, and proposed a new binomial name *Agaricus pseudo-aurantiacus* because of this.

One compound isolated from the fungus is 1,3-diolein, which is in fact an insect attractor. It has been hypothesized that the flies intentionally seek the fly agaric specifically for its intoxicating properties. An alternative derivation proposes that the term *fly-* refers not to insects as such but rather the delirium resulting from consumption of the fungus. This is based on the medieval belief that flies could enter a person's head and cause mental illness. Several regional names appear to be linked with this connotation, meaning either "mad-" or "fool's" *Amanita caesarea.* Hence there is *oriol foll* "mad oriol" in Catalan, *mujolo folo* from Toulouse, *concourlo fouolo* from the Aveyron department in Southern France, *ovolo matto* from Trentino in Italy. A local dialect name in Fribourg in Switzerland is *tsapi de diablhou*, which translates as "Devil's hat".

Classification

Amanita muscaria is the species of the genus. By extension, it is also the type species of *Amanita* subgenus *Amanita*, as well as section *Amanita* within this subgenus. *Amanita* subgenus *Amanita* includes all *Amanita* with inamyloid spores. *Amanita* section *Amanita* includes the species which are very patchy universal veil remnants, including a volva that is reduced to a series of concentric rings and the veil remnants on the cap being a series of patches or warts. Most species in this group also have a bulbous base. *Amanita* section *Amanita* consists of *A. muscaria* and its close relatives, including *A. pantherina* (the panther cap), *A. gemmata*, *A. farinosa*, and *A. xanthocephala.* Modern fungal taxonomists have classified *Amanita muscaria* and its allies this way based on gross morphology and spore inamyloidy. Two recent molecular phylogenetic studies have confirmed this classification as natural.

Figure: *Amanita muscaria var. formosa sensu Thiers, southern Oregon Coast*

Amanita muscaria varies considerably in its morphology and many authorities recognize a number of subspecies or varieties within the species. In *The Agaricales in Modern Taxonomy*, German mycologist Rolf Singer listed three subspecies, though without description: *A. muscaria* ssp. *muscaria*, *A. muscaria* ssp. *americana*, and *A. muscaria* ssp. *flavivolvata*.

Contemporary authorities recognize up to seven varieties:

- var. *muscaria*, the typical red-and-white spotted variety. Some authorities, such as Rodham Tulloss, only use this name for Eurasian and western Alaskan populations.
- var. *flavivolvata* is red, with yellow to yellowish-white warts from southern Alaska down through the Rocky Mountains, through Central America, tall the way to Andean Colombia. Rodham Tulloss uses this name to describe all "typical" *A. muscaria* from indigenous New World populations from Alaska southward.

- var. *alba*, an uncommon fungus, has a white to a silvery white cap that has white warts but is similar to the usual form of mushroom.
- var. *formosa*, has a yellow to orange-yellow cap with yellowish warts and stem.(The stem may be tan). Some authorities utilize the name for all *A. muscaria* which fits this description worldwide (cf. Jenkins), others (cf. Tulloss) restrict its use to Eurasian populations.

Figure: *Amanita muscaria var. guessowii has a yellow to orange cap surface, with the centre of the cap more orange or perhaps even reddish orange.*

Middlesex Fells, Massachusetts

- var. *guessowii* is the color yellow to orange, with the center of cap more orange or perhaps even reddish orange than the outer part of it. It is found most common in northeastern North America, from Newfoundland and Quebec all the way down to the state of Tennessee. Some of the authorities (cf. Jenkins) treat these populations as *A. muscaria* var. *formosa*, while others (cf. Tulloss) recognize it as a distinct variety.
- var. *persicina* is pinkish to orangish and can be called "melon" colored with poorly formed or also some absent remnants of universal veil on the stem and vassal bulb of it, known all the way from the Southeastern Coastal areas of the United States, described in 1977. Recent DNA sequencing suggests this may be a separate species to soon require its own label,
- var. *regalis* which is from Scandinavia and Alaska, is liver-brown and has yellow warts. It appears to be distinctive and

authorities (cf. Tulloss) treat it as a separate species, while others (cf. Jenkins) treat it as a variety of the *A. muscaria*.

A 2006 molecular phylogenetic study of different regional populations of *A. muscaria* by mycologist József Geml and colleagues found three distinct clades within this species representing, roughly, Eurasian, Eurasian "subalpine", and North American populations. Specimens belonging to all three clades have been found in Alaska; this has led to the hypothesis that this was the center of diversification of this species. The study also looked at four named varieties of this species: var. *alba*, var. *flavivolvata*, var. *formosa* (including var. *guessowii*), and var. *regalis* from both areas.

All four varieties were found within both the Eurasian and North American clades, evidence that these morphological forms are simply polymorphisms found throughout the species rather than distinct subspecies or varieties. Further molecular study by Geml and colleagues published in 2008 show these three genetic groups, plus a fourth associated with oak–hickory–pine forest in the southeastern United States, and two more on Santa Cruz Island in California, are delineated from each other enough genetically to be considered separate species; thus *A. muscaria* as it stands currently is a species complex. The complex also includes at least three other closely related taxa currently regarded as species: *A. breckonii* is a buff-capped mushroom associated with conifers from the Pacific Northwest, and the brown-capped *A. gioiosa* and *A. heterochroma* from the mediterranean and Sardinia alone respectively. Both these last two are found with *Eucalyptus* and *Cistus* trees and it is unclear whether they are native or have been introduced from Australia.

Description

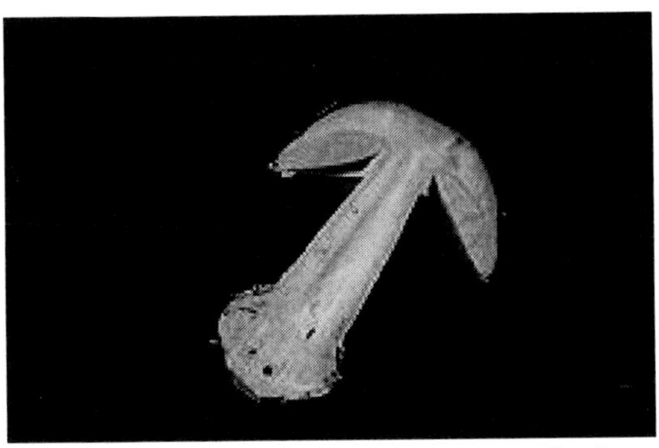

Figure: *Cross section of fruiting body, showing pigment under skin and free gills*

A large conspicuous mushroom, *Amanita muscaria* is generally common and numerous where it grows, and is often found in groups with basidiocarps in all stages of development. Fly agaric fruiting bodies emerge from the soil looking like a white egg, covered in the white warty material of the universal veil. Dissecting the mushroom at this stage will reveal a characteristic yellowish layer of skin under the veil which assists in identification. As the fungus grows, the red color appears through the broken veil and the warts become less prominent; they do not change in size but are reduced relative to the expanding skin area. The cap changes from globose to hemispherical, and finally to plate-like and flat in mature specimens. Fully grown, the bright red cap is usually around 8–20 cm (3–8 in) in diameter, although larger specimens have been found. The red color may fade after rain and in older mushrooms. After emerging from the ground, the cap is covered with numerous small white to yellow pyramid-shaped warts. These are remnants of the universal veil, a membrane that encloses the entire mushroom when it is still very young. The free gills are white, as is the spore print. The oval spores measure 9–13 by 6.5–9 ìm, non-amyloid, and they do not turn blue with the application of iodine. The stipe is white, 5–20 cm high (2–8 in) by 1–2 cm (0.4–0.8 in) wide, and has the slightly brittle, fibrous texture typical of many large mushrooms. At the base is a bulb that bears universal veil remnants in the form of two to four distinct rings or ruffs. Between the basal universal veil remnants and gills are remnants of the partial veil (which covers the gills during development) in the form of a white ring. It can be quite wide and flaccid with age. There is generally no associated smell other than a mild earthiness.

Although very distinctive in appearance, the fly agaric has been mistaken for other yellow to red species in the Americas, like *Armillaria* cf. *mellea* and the edible *Amanita basii*—a Mexican species similar to *A. caesarea* of Europe. Poison control centers in the U.S. and Canada are aware that *amarill* is a common name for *A. caesarea*-like species in Mexico, not just the Spanish for 'yellow'. *Amanita caesarea* can be distinguished as it has an entirely orange to red cap, lacking the numerous white warty spots of the fly agaric. Furthermore the stem, gills and ring are bright yellow, not white. Finally the volva is a distinct white bag, not broken into scales. In Australia, the introduced fly agaric may be confused with the native vermilion grisette (*Amanita xanthocephala*), which grows in association with eucalypts. The latter species generally lacks the white warts of *A. muscaria* and bears no ring.

Distribution and Habitat

Figure: *A. muscaria in a Pinus radiata plantation, near Mount Field National Park, Tasmania*

Amanita muscaria is a cosmopolitan mushroom, native to conifer and deciduous woodlands throughout the temperate and boreal regions of the Northern Hemisphere, includes high elevations of warmer latitudes in regions like Hindu Kush, the Mediterranean and also Central America. A recent molecular study proposes an ancestral origin in the Siberian–Beringian region in the Tertiary period before radiating outwards across all of Asia, Europe and North America. Though generally encountered in autumn, the season can vary in different climates: fruiting occurs in summer and autumn across most of North America, but later in autumn and early winter on the Pacific coast. It is often found in similar locations to *Boletus edulis*, and may appear in fairy rings. Conveyed with pine seedlings, it has been widely transported into the southern hemisphere, including Australia, New Zealand, South Africa and South America, where it can be found in the southern Brazilian state of Paraná.

Ectomycorrhizal, *Amanita muscaria* forms symbiotic relationships with a great variety of trees, including pine, spruce, fir, birch, and cedar. Commonly seen under introduced trees, *A. muscaria* is the fungal equivalent of a weed in New Zealand, Tasmania and Victoria, forming new associations with southern beech (*Nothofagus*). It is also invading a rainforest in Australia, where it may be displacing the native species. Furthermore, it appears to be spreading northwards, with recent reports placing it near Port Macquarie on the New South Wales north coast. Although it has not spread to eucalypts in Australia, it has been recorded associating with them in Portugal.

Toxicity

Amanita muscaria poisoning occurs in either young children or people ingesting it to have a hallucinogenic experience. Occasionally, immature button forms have been mistaken for puffballs. Additionally, the white spots can be washed away during heavy rain and it then may seem as the edible *A. caesarea.*

Figure: *Mature: The white spots may wash off with heavy rainfall*

Amanita muscaria contains a number of biologically active agents, at least one of which, muscimol, is known to be psychoactive. Ibotenic acid, a neurotoxin, serves as a prodrug to muscimol, with approximately 10-20% converting to muscimol upon ingestion. A toxic dose in adults is approximately 6 mg muscimol or 30 to 60 mg ibotenic acid; this is typically about the amount found in one cap of *Amanita muscaria.* However, the amount and ratio of chemical compounds per mushroom varies widely from region to region and season to season, which further confuses the issue. Spring and summer mushrooms have been reported to contain up to 10 times as much ibotenic acid and muscimol compared to autumn fruitings.

A fatal dose has been calculated at an amount of 15 caps. Deaths from this fungus *A. muscaria* have been reported in historical journal articles and newspaper reports; however, with modern medical treatment a fatal outcome because of the poison of this mushroom would be extremely rare. Many older books list it as "deadly" but this is a mistake that gives the impression it is far more toxic than it actually is. The North American Mycological Association has stated there are absolutely no reliably documented fatalities in the past century. The vast majority (90% or more) of mushroom poisoning deaths are from having eaten either the greenish to yellowish death cap (*A. phalloides*) or perhaps even one of the several white *Amanita* species which are known as destroying angels.

The active constituents of this species are water soluble, and boiling and then discarding the cooking water will at least partly detoxify *A. muscaria*. However, drying may increase potency as the process facilitates the conversion of ibotenic acid to the more potent muscimol. According to some sources, once detoxified, the mushroom becomes edible.

Pharmacology

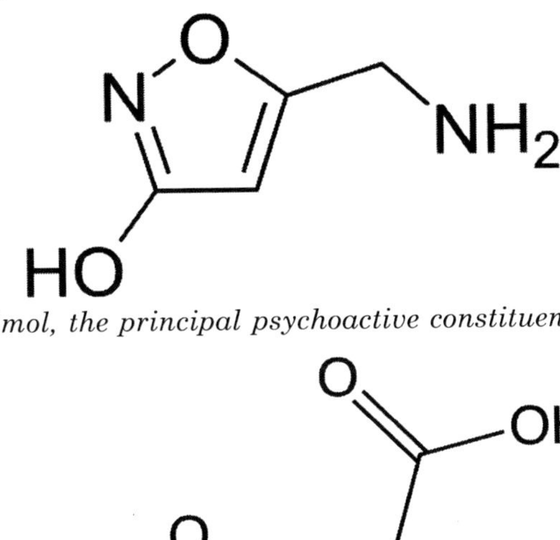

Figure: *Muscimol, the principal psychoactive constituent of* A. muscaria.

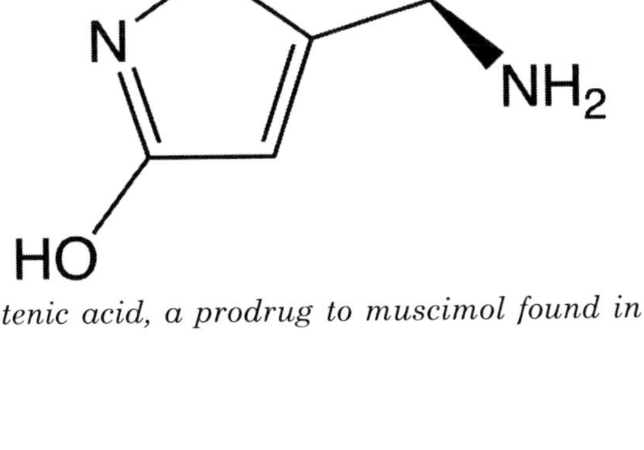

Figure: *Ibotenic acid, a prodrug to muscimol found in* A. muscaria.

Muscarine, discovered in 1869, was long thought to be the active hallucinogenic agent in *A. muscaria*. Muscarine binds with muscarinic acetylcholine receptors leading to the excitation of neurons bearing these receptors. The levels in *Amanita muscaria*, however, are minute when compared with other poisonous fungi, such as *Inocybe erubescens* or small white *Clitocybe* species *C. dealbata* and *C. rivulosa*, and are too insignificant to play a role in the symptoms of poisoning.

The major toxins involved in poisoning are muscimol (3-hydroxy-5-aminomethyl-1-isoxazole, an unsaturated cyclic hydroxamic acid) and the related amino-acid ibotenic acid. Muscimol is the product of the decarboxylation (usually by drying) of ibotenic acid. Muscimol and ibotenic acid were discovered in the mid-20th century. Researchers in England, Japan, and Switzerland showed that the effects produced were due mainly to ibotenic acid and muscimol, not muscarine. These toxins are not distributed uniformly in the mushroom. Most are detected in the cap of the fruit, rather than in the base, with the smallest amount in the stalk. (Lampe, 1978; Tsunoda et al., 1993) A substantial fraction of ingested ibotenic acid is excreted in the urine unmetabolized quite rapidly, between 20 and 90 minutes after ingestion. Virtually no muscimol is excreted when pure ibotenic acid is eaten but muscimol is detectable in the urine after eating *A. muscaria*, which contains both ibotenic acid and muscimol.

Ibotenic acid and muscimol are structurally related to each other and to two major neurotransmitters of the central nervous system: glutamic acid and GABA respectively. Ibotenic acid and muscimol act like these neurotransmitters, muscimol being a potent $GABA_A$ agonist, while ibotenic acid is an agonist of NMDA glutamate receptors and certain metabotropic glutamate receptors which are involved in the control of neuronal activity. It is these interactions which are thought to cause the psychoactive effects found in intoxication. Muscimol is the agent responsible for the majority of the psychoactivity.

Muscazone is another compound more recently isolated from European specimens of the fly agaric. It is a product of the breakdown of ibotenic acid by ultra-violet radiation. Muscazone is of minor pharmacological activity compared with the other agents. *Amanita muscaria* and related species are known as effective bioaccumulators of vanadium; some species concentrate vanadium to levels of up to 400 times those typically found in plants. Vanadium is present in fruit-bodies as an organometallic compound called amavadine. However, the biological importance of the accumulation process is unknown.

Symptoms

Fly agarics are known for the unpredictability of their effects. Depending on habitat and the amount ingested per body weight, effects can range from nausea and twitching to drowsiness, cholinergic crisis-like effects (low blood pressure, sweating and salivation), auditory and visual distortions, mood changes, euphoria, relaxation, ataxia, and loss of equilibrium.

In cases of serious poisoning it causes a delirium, somewhat similar in effect to anticholinergic poisoning (such as that caused by *Datura stramonium*), characterized by bouts of marked agitation with confusion, hallucinations, and irritability followed by periods of central nervous system depression. Seizures and coma may also occur in severe poisonings. Symptoms typically appear after around 30 to 90 minutes and peak within three hours, but certain effects can last for a number of days. In the majority of cases recovery is complete within 12 to 24 hours. The effect is highly variable between individuals with similar doses potentially causing quite different reactions. Some cases of intoxication have exhibited headaches up to ten hours afterwards. Retrograde amnesia and somnolence can result following recovery.

Treatment

Medical attention should be sought in cases of suspected poisoning. Initial treatment consists of gastric decontamination. If the delay between ingestion and treatment is less than four hours, activated charcoal is given. Gastric lavage can be considered if the patient presents within 1 hour of ingestion. Inducing vomiting with syrup of ipecac is no longer recommended in any poisoning situations.

There is no antidote, and supportive care is the mainstay of further treatment for intoxication. Patients can develop psychological symptoms similar to anticholinergic poisoning and physical symptoms similar to cholinergic poisoning; however, the use of atropine or physostigmine as an antidote is not recommended as muscimol and ibotenic acid do not produce a true anticholinergic syndrome nor do they have activity at muscarinic receptors. If a patient is delirious or agitated, this can usually be treated by reassurance and, if necessary, physical restraints. Additionally, benzodiazepine such as diazepam or lorazepam can be used to control combativeness, agitation, muscular overactivity, and seizures. However, only small doses of benzodiazepines should be used as they may worsen the respiratory depressant effects of muscimol. Recurrent vomiting is rare but if present may lead to fluid and electrolyte imbalances; intravenous rehydration or electrolyte replacement may be required. Serious cases may develop loss of consciousness or coma, and may necessitate intubation and artificial ventilation. Hemodialysis can remove the toxins, although this intervention is generally considered unnecessary. With modern medical treatment the prognosis is typically good following supportive treatment.

Psychoactive Use

Unlike psilocybin mushrooms, *Amanita muscaria* has been rarely consumed because of its toxicity and unpredictable psychological effects. However, following the outlawing of psilocybin mushrooms in the United Kingdom, an increased quantity of legal *A. muscaria* mushrooms began to be sold for recreational and entheogenic use. Professor Marija Gimbutas, a renowned Lithuanian historian, reported to R. Gordon Wasson on the use of this mushroom in Lithuania. In remote areas of Lithuania *Amanita muscaria* has been taken at wedding feasts, in which mushrooms were mixed with vodka. The professor also reported that the Lithuanians used to export quantities of *A. muscaria* to the Lapps in the Far North for use in shamanic rituals. The Lithuanian festivities are the only report that Wasson received of ingestion of fly agaric for recreational use in Eastern Europe.

Siberia

Amanita muscaria was widely used as an entheogen by many of the indigenous peoples of Siberia. Its use was known among almost all of the Uralic-speaking peoples of western Siberia and the Paleosiberian-speaking peoples of the Russian Far East. However, there are only isolated reports of *A. muscaria* use among the Tungusic and Turkic peoples of central Siberia and it is believed that entheogenic use of *A. muscaria* was largely not a practice of these peoples. In western Siberia, the use of *A. muscaria* was restricted to shamans, who used it as an alternate method of achieving a trance state. (Normally, Siberian shamans achieve a trance state by prolonged drumming and dancing.) In eastern Siberia, *A. muscaria* was used by both shamans and laypeople alike, and was used recreationally as well as religiously. In eastern Siberia, the shaman would consume the mushrooms, and others would drink his urine. This urine, still containing psychoactive elements, may actually be more potent than the *A. muscaria* mushrooms with fewer negative effects, such as sweating and twitching, suggesting that the initial user may act as a screening filter for other components in the mushroom.

The Koryak of eastern Siberia have a story about the fly agaric (*wapaq*) which enabled Big Raven to carry a whale to its home. In the story, the deity *Vahiyinin* ("Existence") spat onto earth, and his spittle became the *wapaq*, and his saliva becomes the warts. After experiencing the power of the *wapaq*, Raven was so exhilarated that he told it to grow forever on earth so his children, the people, could learn from it. Among the Koryak, one report held the poor would consume the urine of the wealthy, who could afford to buy the mushrooms.

Other Reports of Entheogenic Use

Beyond Siberia, there are only isolated and unconfirmed reports of the entheogenic use of *A. muscaria*. The Finnish historian T. I. Itkonen mentions that it was once used among the Sami people, sorcerers in Inari would consume fly agarics with seven spots. In 1979, Said Gholam Mochtar and Hartmut Geerken published an article in which they claim to have discovered a tradition of medicinal and recreational use of this mushroom among a Parachi-speaking group in Afghanistan. There are also unconfirmed reports of religious use of *A. muscaria* among two Subarctic Native American tribes. Ojibwa ethnobotanist Keewaydinoquay Peschel reported its use among her people, where it was known as the *miskwedo*. This information was enthusiastically received by Wasson, although evidence from ther sources was lacking. There is also one account of a Euro-American who claims to have been initiated into traditional Tlicho use of *Amanita muscaria*.

Soma

In 1968, R. Gordon Wasson proposed that *A. muscaria* was the *Soma* talked about in the Rig Veda of India, which received widespread publicity and popular support at the time. He noted that descriptions of *Soma* omitted description of roots, stems or seeds, which suggested a mushroom, and used the adjective *hári* "dazzling" or "flaming" which the author interprets as red. One line described men urinating *Soma*; this recalled the practice of recycling urine in Siberia.

Soma is mentioned as coming "from the mountains", which Wasson interpreted as being brought with the Aryan invaders from the north. However, Indian scholars Santosh Kumar Dash and Sachinanda Padhy noted that both the eating of mushrooms and drinking of urine were proscribed, using as a source the *Manusm[ti*. In 1971, Vedic scholar John Brough from Cambridge University rejected Wasson's theory; he noted that the language was too vague to determine a description of Soma. In his 1976 survey, *Hallucinogens and Culture*, anthropologist Peter T. Furst evaluated the evidence for and against the identification of the fly agaric mushroom as Vedic Soma, concluding cautiously in its favor.

Vikings

A single source for the notion that Vikings used *A. muscaria* to produce their berserker rages was first suggested by the Swedish professor Samuel Ödmann in 1784. Ödmann based his theories on reports about the use of fly agaric among Siberian shamans. The notion has become widespread since the 19th century, but no contemporary sources mention this use or anything similar in their description of berserkers. Today, it is generally considered an urban legend or at best speculation that cannot be proven. Muscimol is generally a mild relaxant, but could create a range of reactions within a range of people. It is possible that it could make a person incredibly angry, as well as make them "very jolly or sad, jump about, dance, sing or give way to great fright".

Christianity

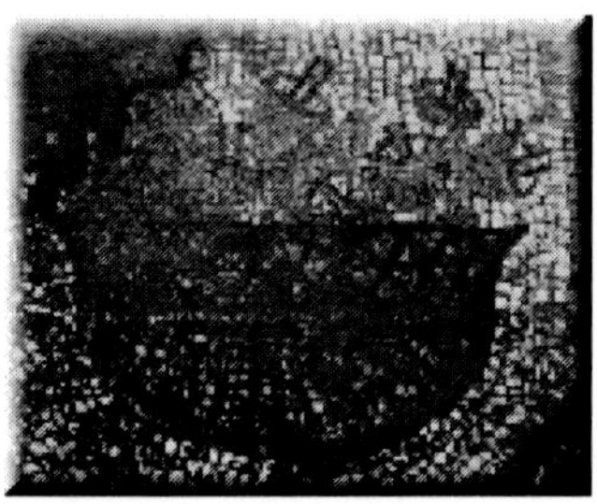

Figure: *Mosaic of red mushrooms, found in the Christian Basilica of Aquileia in northern Italy, dating to before 330 AD.*

Biblical scholar John Marco Allegro controversially proposed that the Roman Theology was derived from a sex and psychedelic mushroom cult in his 1970 book *The Sacred Mushroom and the Cross*, although his theory has found little support by scholars outside the field of ethnomycology. The book was roundly discredited by academics and theologians, including Sir Godfrey Driver, Emeritus Professor of Semitic Philology at Oxford University, and Henry Chadwick, the Dean of Christ Church College, Oxford. Christian author John C. King wrote a detailed rebuttal of Allegro's theory in the 1970 book *A Christian View of the Mushroom Myth*; he notes neither fly agarics nor their host trees are found in the middle east even though cedars and pines are found there, and highlights the tenuous nature of the links between biblical and Sumerian names coined by Allegro. He concludes that if the theory was true, the use of the mushroom must have been "the best kept secret in the world" as it was so well concealed for all this time.

In *Magic Mushrooms in Religion and Alchemy* (formerly called *Strange Fruit*), Clark Heinrich interprets *A. muscaria* usage by Adam and Eve, Moses, Elijah and Elisha, Isaiah, Ezekiel, Jonah, Jesus and his disciples, and John of Patmos. In the book *Apples of Apollo* the mushroom is identified in a wide range of mythological tales such as those involving Perseus, Prometheus, Heracles, Jason and the Argonauts, Jesus and the Holy Grail.

Culinary Use

The toxins in *A. muscaria* are water soluble. When sliced thinly, or chopped into thin dice and boiled in plentiful water until thoroughly cooked, it seems to be detoxified. Although its consumption as a food has never been widespread, the consumption of detoxified *A. muscaria* has been practiced in some localities in Europe (notably by Russian settlers in Siberia) since at least the 19th century, and likely earlier. The German physician and naturalist Georg Heinrich von Langsdorff wrote the earliest published account on how to detoxify this mushroom in 1823. In the late 19th Century, the French physician Félix Archimède Pouchet was a popularizer and advocate of *A. muscaria* consumption, comparing it to manioc, an important food source in tropical South America that nevertheless must be detoxified before consumption.

Use of this mushroom as a food source also seems to have existed in North America as well. A classic description of this use of *A. muscaria* by an African-American mushroom seller in Washington, D.C., in the late 19th century is described by American botanist Frederick Vernon Coville. In this case, the mushroom, after parboiling, and soaking in vinegar is made into a mushroom sauce for steak. It is also consumed as a food in parts of Japan. The most well-known current use as an edible mushroom is in Nagano Prefecture, Japan. There, it is primarily salted and pickled.

A 2008 paper by food historian William Rubel and mycologist David Arora gives a history of consumption of *A. muscaria* as a food and describes detoxification methods. They advocate that *Amanita muscaria* be described in field guides as an edible mushroom, though accompanied by a description on how to detoxify it. The authors state that the widespread descriptions in field guides of this mushroom as poisonous is a reflection of cultural bias, as several other popular edible species, notably morels, are toxic unless properly cooked.

Cultural Depictions

The red-and-white spotted toadstool is a common image in many aspects of popular culture, especially in children's books, film, garden ornaments, greeting cards, and more recently computer games. Garden ornaments, and children's picture books depicting gnomes and fairies, such as the Smurfs, very often show fly agarics used as seats, or homes. Fly agarics have been featured in paintings since the Renaissance, albeit in a subtle manner. In the Victorian era they became more visible, even becoming the main topic of some fairy paintings. Two of the most famous uses of the mushroom are in the video game series *Super Mario Bros.*, and the dancing mushroom sequence in the 1940 Disney film *Fantasia*.

Figure: *Moritz von Schwind's 1851 painting of Rübezahl features fly agarics.*

Literature

The journeys of Philip von Strahlenberg to Siberia and his descriptions of the use of the *mukhomor* there was published in English in 1736. The drinking of urine of those who had imbibed the mushroom was commented on by Anglo-Irish writer Oliver Goldsmith in his widely read 1762 novel *Citizen of the World*. The mushroom had been identified as the fly agaric by this time. Other authors recorded the distortions of the size of perceived objects while intoxicated by the fungus, including naturalist Mordecai Cubitt Cooke in his books *The Seven Sisters of Sleep* and *A Plain and Easy Account of British Fungi*. This observation is thought to have formed the basis of the effects of eating the mushroom in the 1865 popular story *Alice's Adventures in Wonderland*. A hallucinogenic "scarlet toadstool" from Lappland is also featured as a plot element in Charles Kingsley's 1866 novel *Hereward the Wake* based on the medieval figure of the same name; fly agaric shamanism is explored more recently in the 2003 novel *Thursbitch* by Alan Garner.

Figure: *Children play on Jose de Creeft's sculpture Alice in Wonderland in Eastern Central Park, New York. Alice sits on a mushroom, inviting children to climb up and join her. The mushroom in the sculpture is not a faithfully reproduced Amanita muscaria; the reference within Lewis Carroll's original literary work upon which the sculpture is based is often discussed.*

Christmas Decorations and Santa Claus

Fly agarics appear on Christmas cards and New Year cards from around the world as a symbol of good luck. The ethnobotanist Jonathan Ott has suggested that the idea of Santa Claus and tradition of hanging stockings over the fireplace is based centrally upon the fly agaric mushroom itself. With its generally red and white color scheme, he argues that Santa Claus's suit is related to the mushroom. He also draws parallels with flying reindeer: reindeer had been reported to consume the mushroom and prance around in an intoxicated manner afterwards. American ethnopharmacologist Scott Hajicek-Dobberstein, researching possible links between religious myths and the red mushroom, notes, "If Santa Claus had but one eye [like Odin], or if magic urine had been a part of his legend, his connection to the *Amanita muscaria* would be much easier to believe.".

The connection was reported to a much wider audience with an article in the magazine of *The Sunday Times* in 1980, and *New Scientist* in 1986. Historian Ronald Hutton has since disputed the connection; he noted reindeer spirits did not appear in Siberian mythology, shamans did not travel by sleigh, nor did they wear red and white, or climb out of smoke holes in yurt roofs. Finally, American awareness of Siberian shamanism postdated the appearance of much of the folklore around Santa.

Amanita Pantherina var. Pantherina

Amanita pantherina var. pantherina, also known as the panther cap and false blusher due to its similarity to the true blusher *(Amanita rubescens)*, is a species of Europe and western Asia.

Description

- *Cap:* 4 – 11 cm wide, hemispheric at first, then convex to plano-convex, deep brown to hazel-brown to pale ochraceous brown, densely distributed warts that are pure white to sordid cream, minutely verruculose, floccose, easily removable. Viscid when wet, with a short striate margin. The flesh is white, unchanging when injured.
- *Gills:* free, close to crowded, white becoming greyish, truncate.
- *Spores:* white in deposit, broadly ellipsoid to ellipsoid to elongate, infrequently globose. 8 — 12 × 5.5 — 8 μm.
- *Stipe:* 5 – 14 cm long × .6 – 2 cm wide, subcyclindric, somewhat narrowing upward, white, becoming slightly tannish in age, stuffed then hollow, finely floccose becoming smooth above the ring, and with small appressed squamules or creamy floccose material below. The volva is white, becoming grey with age, forming one or sometimes two narrow hoop-like rings just above the bulbous base. The flesh is white, unchanging when injured.
- *Odour:* Unpleasant or like raw potatoes
- *Microscopic features:* Spores are 8-14 x 6-10 μm, smooth, elliptical and inamyloid.

Other than the brownish cap with white warts, distinguishing features of *Amanita pantherina* include the collar-like roll of volval tissue at the top of the basal bulb, and the elliptical, inamyloid spores.

Habitat and Distribution

The panther cap is an uncommon mushroom, found in both deciduous, especially beech and, less frequently, coniferous woodland and rarely meadows throughout Europe, western Asia in late summer and autumn. It has also been recorded from South Africa, where it is thought to have been accidentally introduced with trees imported from Europe.

It is an ectomycorrhizal fungus, living in root symbiosis with a tree, deriving photosynthesised nutrients from it and providing soil nutrients in return.

Biochemistry

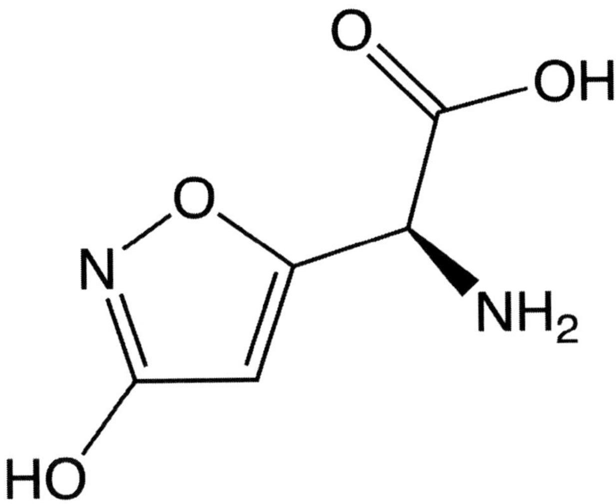

Figure: *Ibotenic acid*

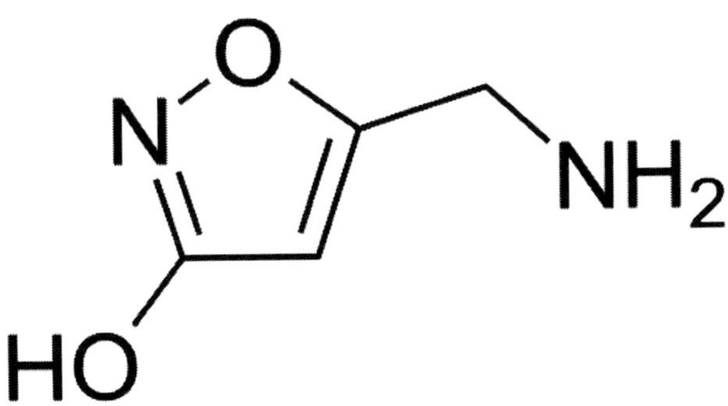

Figure: *Muscimol*

Amanita pantherina contains the psychoactive compound muscimol, but is used as an entheogen much less often than the related

Amanita muscaria because of the potentially dangerously higher levels of muscimol found in the mushroom.

Amanita Ocreata

Amanita ocreata, commonly known as the death angel, destroying angel, angel of death or more precisely Western North American destroying angel, is a deadly poisonous basidiomycete fungus, one of many in the genus *Amanita*. Occurring in the Pacific Northwest and California floristic provinces of North America, *A. ocreata* associates with oak trees. The large fruiting bodies (the mushrooms) generally appear in spring; the cap may be white or ochre and often develops a brownish centre, while the stipe, ring, gill and volva are all white.

Amanita ocreata resemble several edible species commonly consumed by humans, increasing the risk of accidental poisoning. Mature fruiting bodies can be confused with the edible *A. velosa*, *A. lanei* or *Volvariella speciosa*, while immature specimens may be difficult to distinguish from edible *Agaricus* mushrooms or puffballs.

Similar in toxicity to the death cap (*A. phalloides*) and destroying angels of Europe (*A. virosa*) and eastern North America (*A. bisporigera*), it is a potentially deadly fungus responsible for a number of poisonings in California. Its principal toxic constituent, á-amanitin, damages the liver and kidneys, often fatally, and has no known antidote. The initial symptoms are gastrointestinal and include colicky abdominal pain, diarrhea and vomiting.

These subside temporarily after 2–3 days, though ongoing damage to internal organs during this time is common; symptoms of jaundice, diarrhea, delirium, seizures, and coma may follow with death from liver failure 6–16 days post ingestion.

Taxonomy and Naming

Amanita ocreata was first described by American mycologist Charles Horton Peck in 1909 from material collected by Charles Fuller Baker in Claremont, California. The specific epithet is derived from the Latin *ocrìâtus* 'wearing greaves' from *ocrea* 'greave', referring to its loose, baggy volva. *Amanita bivolvata* is a botanical synonym. The mushroom belongs to the same section (*Phalloideae*) and genus (*Amanita*) as several deadly poisonous fungi including the death cap (*A. phalloides*) and several all-white species of *Amanita* known as "destroying angels": *A. bisporigera* of eastern North America, and the European *A. virosa*. "Death angel" is used as an alternate common name.

Description

A. ocreata is generally stouter than the other fungi termed destroying angels. It first appears as a white egg-shaped object covered with a universal veil. As it grows, the mushroom breaks free, though there may rarely be ragged patches of veil left at the cap edges. The cap is initially hemispherical, before becoming more convex and flattening, sometimes irregularly. This may result in undulations in the cap, which may reach up to 12 cm (5 in) in diameter. The colour varies from white, through yellowish-white to shades of ochre, sometimes with a brownish centre. Occasionally parts of the fruiting bodies may have pinkish tones.

The rest of the fungus below the cap is white. The crowded gills are free to narrowly adnate. The stipe is 8–20 cm (3–8 in) high and 1.5–2 cm (½–T! in) thick at the apex, and bears a thin white membranous ring. The volva is thin, smooth and sac-like, although may be quite extensive and contain almost half the stipe. The spore print is white, and the subglobose to ovoid to subellipsoid, amyloid spores are 9–14 x 7–10 ìm viewed under a microscope. There is typically no smell, though some fruiting bodies may have a slight odour, described as that of bleach or chlorine, dead fish or iodine. Like other destroying angels, the flesh stains yellow when treated with potassium hydroxide (KOH).

This fungus resembles the edible mushrooms *Agaricus arvensis* and *A. campestris*, and the puffballs (*Lycoperdon* spp.) before the caps have opened and the gills have become visible, so those collecting immature fungi run the risk of confusing the varieties. It also resembles and grows in the same areas as the edible and prized *Amanita velosa*, which can be distinguished from *A. ocreata* by its lack of ring, striate cap margin and thick universal veil remnants comprising the veil. The edible *Amanita lanei* lacks a ring and is more likely to have veil patches remaining on its cap, which is generally darker. *Volvariella speciosa* has pink spores and no ring or volva.

Distribution and Habitat

Appearing from January to April, *A. ocreata* occurs later in the year than other amanitas except *A. calyptroderma*. It is found in mixed woodland on the Pacific coast of North America, from Washington south through California to Baja California in Mexico. It may feasibly occur on Vancouver Island in British Columbia though this has never been confirmed. It forms ectomycorrhizal relationships and is found in association with coast live oak (*Quercus agrifolia*), as well as hazel

(*Corylus* spp.). In Oregon and Washington, it may also be associated with the Garry oak (*Quercus garryana*).

Toxicity

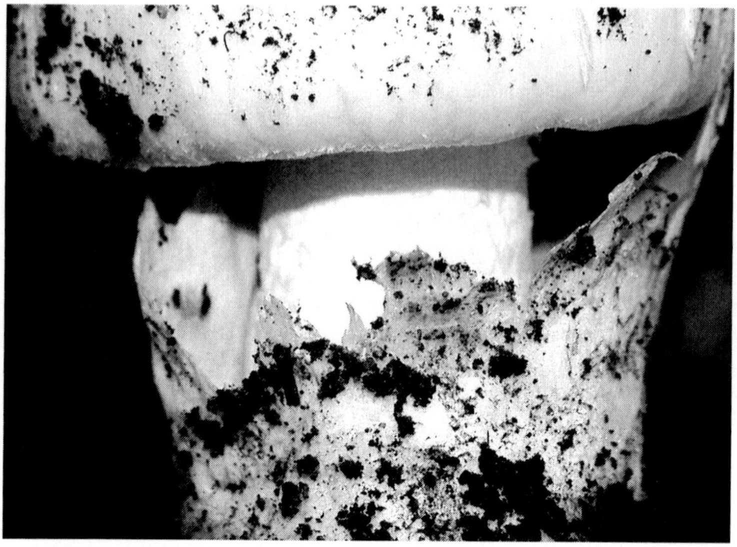

***Figure:** Mushroom emerging through universal veil,*

Petaluma, California

Amanita ocreata is highly toxic, and has been responsible for a number of mushroom poisonings in western North America, particularly in the spring. It contains highly toxic amatoxins, as well as phallotoxins, a feature shared with the closely related death cap (*A. phalloides*), half a cap of which can be enough to kill a human, and other species known as destroying angels. There is some evidence it may be the most toxic of all the North American phalloideae, as a higher proportion of people consuming it had organ damage and 40% perished. Dogs, too, have been known to consume this fungus in California with fatal results.

Amatoxins consist of at least eight compounds with a similar structure, that of eight amino-acid rings; of those found in *A. ocreata,* á-amanitin is the most prevalent and along with â-amanitin is likely to be responsible for the toxic effects. The major toxic mechanism is the inhibition of RNA polymerase II, a vital enzyme in the synthesis of messenger RNA (mRNA), microRNA, and small nuclear RNA (snRNA). Without mRNA, essential protein synthesis and hence cell metabolism stop and the cell dies. The liver is the principal organ affected, as it is the first organ encountered after absorption by the gastrointestinal tract, though other organs, especially the kidneys, are susceptible to the toxins. The phallotoxins consist of at least seven compounds, all of which have seven similar peptide rings. Although they are highly toxic to liver cells, phallotoxins have since been found to have little input into the destroying angel's toxicity as they are not absorbed through the gut. Furthermore, one phallotoxin, phalloidin, is also found in the edible (and sought-after) blusher (*Amanita rubescens*).

Symptoms

Symptoms of poisoning by *A. ocreata* are initially gastrointestinal in nature and include colicky abdominal pain, with watery diarrhea and vomiting which may lead to dehydration, and, in severe cases, hypotension, tachycardia, hypoglycemia, and acid-base disturbances. The initial symptoms resolve two to three days after ingestion of the fungus. A more serious deterioration signifying liver involvement may then occur—jaundice, diarrhea, delirium, seizures, and coma due to fulminant hepatic failure and attendant hepatic encephalopathy caused by the accumulation of normally liver-removed substances in the blood. Renal failure (either secondary to severe hepatitis or caused by direct toxic renal damage) and coagulopathy may appear during this stage. Life-threatening complications include increased intracranial pressure, intracranial hemorrhage, sepsis, pancreatitis, acute renal failure, and cardiac arrest. Death generally occurs six to sixteen days after the poisoning.

Treatment

Consumption of *A. ocreata* is a medical emergency that requires hospitalization. There are four main categories of therapy for poisoning: preliminary medical care, supportive measures, specific treatments, and liver transplantation.

Preliminary care consists of gastric decontamination with either activated carbon or gastric lavage. However, due to the delay between ingestion and the first symptoms of poisoning, it is commonplace for patients to arrive for treatment long after ingestion, potentially reducing the efficacy of these interventions. Supportive measures are directed towards treating the dehydration which results from fluid loss during the gastrointestinal phase of intoxication and correction of metabolic acidosis, hypoglycemia, electrolyte imbalances, and impaired coagulation.

No definitive antidote for amatoxin poisoning is available, but some specific treatments such as intravenous penicillin G have been shown to improve survivability. There is some evidence that intravenous silibinin, an extract from the blessed milk thistle (*Silybum marianum*), may be beneficial in reducing the effects of amatoxins, preventing their uptake by hepatocytes, thereby protecting undamaged hepatic tissue. In patients developing liver failure, a liver transplant is often the only option to prevent death. Liver transplants have become a well-established option in amatoxin poisoning. This is a complicated issue, however, as transplants themselves may have significant complications and mortality; patients require long-term immunosuppression to maintain the transplant. Evidence suggests that, although survival rates have improved with modern medical treatment, in patients with moderate to severe poisoning up to half of those who did recover suffered permanent liver damage. However, a follow-up study has shown that most survivors recover completely without any sequelae if treated within 36 hours of mushroom ingestion.

Amanita Phalloides

Amanita phalloides (pron.: /æmYÈnajtY fYÈlTjdiÐz/), commonly known as the death cap, is a deadly poisonous basidiomycete fungus, one of many in the genus *Amanita*. Widely distributed across Europe, *A. phalloides* forms ectomycorrhizas with various broadleaved trees. In some cases, death cap has been introduced to new regions with the cultivation of non-native species of oak, chestnut, and pine. The large fruiting bodies (mushrooms) appear in summer and autumn; the caps are generally greenish in color, with a white stipe and gills.

Coincidentally, these toxic mushrooms resemble several edible species (most notably caesar's mushroom and the straw mushroom) commonly consumed by humans, increasing the risk of accidental poisoning. *A. phalloides* is one of the most poisonous of all known toadstools. It has been involved in the majority of human deaths from mushroom poisoning, possibly including the deaths of Roman Emperor Claudius and Holy Roman Emperor Charles VI. It has been the subject of much research, and many of its biologically active agents have been isolated. The principal toxic constituent is á-amanitin, which damages the liver and kidneys, often fatally.

Taxonomy and Naming

The death cap was first described by French botanist Sébastien Vaillant in 1727, who gave a succinct phrase name "*Fungus phalloides, annulatus, sordide virescens, et patulus*", which is still recognizable as the fungus today. Though the scientific name *phalloides* means "phallus-shaped", it is unclear whether it is named for its resemblance to a literal phallus or the stinkhorn mushrooms *Phallus*. In 1821, Elias Magnus Fries described it as *Agaricus phalloides*, but included all white amanitas within its description.

Finally in 1833, Johann Heinrich Friedrich Link settled on the name *Amanita phalloides*, after Persoon had named it *Amanita viridis* 30 years earlier. Although Louis Secretan's use of the name *Amanita phalloides* predates Link's, it has been rejected for nomenclatural purposes because Secretan's works did not use binomial nomenclature consistently; some taxonomists have, however, disagreed with this opinion.

A. phalloides is the type species of *Amanita* section Phalloideae, a group that contains all of the deadly poisonous *Amanita* species thus far identified. Most notable of these are the species known as destroying angels, namely *Amanita virosa* and *Amanita bisporigera*, as well as the fool's mushroom *(A. verna)*. The term "destroying angel" has been applied to *A. phalloides* at times, but "death cap" is by far the most common vernacular name used in English. Other common names also listed include "stinking amanita" and "deadly amanita".

A rarely appearing, all-white form was initially described *A. phalloides* f. *alba* by Max Britzelmayr, though its status has been unclear. It is often found growing amid normally colored death caps. It has been described, in 2004, as a distinct variety and includes what was termed *A. verna* var. *tarda*. The true *Amanita verna* fruits in spring and turns yellow with KOH solution, whereas *A. phalloides* never does.

Description

The death cap has a large and imposing epigeous (aboveground) fruiting body (basidiocarp), usually with a pileus (cap) from 5 to 15 cm (2–6 in) across, initially rounded and hemispherical, but flattening with age. The color of the cap can be pale-, yellowish-, or olive-green, often paler toward the margins and often paler after rain. The cap surface is sticky when wet and easily peeled, a troublesome feature, as that is allegedly a feature of edible fungi. The remains of the partial veil are seen as a skirtlike, floppy annulus usually about 1.0 to 1.5 cm (0.4–0.6 in) below the cap. The crowded white lamellae (gills) are free. The stipe is white with a scattering of grayish-olive scales and is 8 to 15 cm (3–6 in) long and 1 to 2 cm (3/8–3/4 in) thick, with a swollen, ragged, sac-like white volva (base). As the volva, which may be hidden by leaf litter, is a distinctive and diagnostic feature, it is important to remove some debris to check for it.

The smell has been described as initially faint and honey-sweet, but strengthening over time to become overpowering, sickly-sweet and objectionable. Young specimens first emerge from the ground resembling a white egg covered by a universal veil, which then breaks, leaving the volva as a remnant. The spore print is white, a common feature of *Amanita*. The transparent spores are globular to egg-shaped, measure 8–10 ìm (0.3–0.4 mil) long, and stain blue with iodine. The gills, in contrast, stain pallid lilac or pink with concentrated sulfuric acid.

Distribution and Habitat

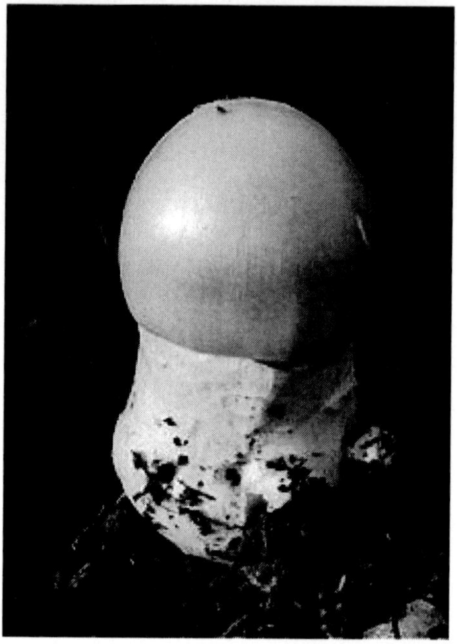

Figure: *A young death cap emerging from its universal veil*

The death cap is native to Europe, where it is widespread. It is found from the southern coastal regions of Scandinavia in the north, to Ireland in the west, east to Poland and western Russia, and south throughout the Balkans, in Italy, Spain and Portugal, and in Morocco and Algeria in north Africa. In west Asia it has been reported from forests of northern Iran. There are records from further east into Asia but these have yet to be confirmed as *A. phalloides*.

It is ectomycorrhizally associated with a number of tree species and is symbiotic with them. In Europe, these include a large number of hardwood and, less frequently, conifer species. It appears most commonly under oaks, but also under beeches, chestnuts, horse-chestnuts, birches, filberts, hornbeams, pines, and spruces. In other areas, *A. phalloides* may also be associated with these trees or with only some species and not others. In coastal California, for example, *A. phalloides* is associated with coast live oak, but not with the various coastal pine species, such as Monterey pine. In countries where it has been introduced, it has been restricted to those exotic trees with which it would associate in its natural range. There is, however, evidence of *A. phalloides* associating with hemlock and with genera of the Myrtaceae: *Eucalyptus* in Tanzania and Algeria, and *Leptospermum* and *Kunzea* in New Zealand. This suggests the species may have invasive potential.

By the end of the 19th century, Charles Horton Peck had reported *A. phalloides* in North America. However, in 1918, samples from the eastern United States were identified as being a distinct though similar species, *A. brunnescens*, by G. F. Atkinson of Cornell University. By the 1970s, it had become clear that *A. phalloides* actually does occur in the United States, apparently having been introduced from Europe alongside chestnuts, with populations on the West and East Coasts. Although a 2006 historical review concluded the East Coast populations were introduced, the origins of the West Coast populations remained unclear, owing to the scantiness of historical records. However, a 2009 genetic study provided strong evidence for the introduced status of the fungus on the west coast of North America.

Amanita phalloides has been conveyed to new countries across the Southern Hemisphere with the importation of hardwoods and conifers. Introduced oaks appear to have been the vector to Australia and South America; populations under oaks have been recorded from Melbourne and Canberra (where two people died in January 2012, of four who were poisoned) and Adelaide, as well as Uruguay. It has been recorded under other introduced trees in Argentina and Chile. Pine plantations are associated with the fungus in Tanzania and South Africa, where it is also found under oaks and poplars.

Toxicity

As the common name suggests, the fungus is highly toxic, and is responsible for the majority of fatal mushroom poisonings worldwide. Its biochemistry has been researched intensively for decades, and 30 grams (1 oz), or half a cap, of this mushroom is estimated to be enough to kill a human. In 2006, a family of three in Poland was poisoned, resulting in one death and the two survivors requiring liver transplants. Some authorities strongly advise against putting suspected death caps in the same basket with fungi collected for the table and to avoid touching them. Furthermore, the toxicity is not reduced by cooking, freezing, or drying.

Similarity to Edible Species

In general, poisoning incidents are unintentional and result from errors in identification. Recent cases highlight the issue of the similarity of *A. phalloides* to the edible paddy straw mushroom, *Volvariella volvacea*, with East- and Southeast-Asian immigrants in Australia and the west coast of the United States falling victim. In an episode in Oregon, four members of a Korean family required liver transplants. Of the 9 people poisoned in the Canberra region between 1988 and 2011, three were from Laos and two were from China. This misidentification is a leading cause of mushroom poisoning in the United States.

Novices may mistake juvenile death caps for edible puffballs or mature specimens for other edible *Amanita* species, such as *Amanita lanei*, so some authorities recommend avoiding the collecting of *Amanita* species for the table altogether. The white form of *A. phalloides* may be mistaken for edible species of *Agaricus*, especially the young fruitbodies whose unexpanded caps conceal the telltale white gills; all mature species of *Agaricus* have dark-colored gills.

In Europe, other similarly green-capped species collected by mushroom hunters include various green-hued brittlegills of the genus *Russula* and the formerly popular *Tricholoma flavovirens*, now regarded as hazardous owing to a series of restaurant poisonings in France. Brittlegills, such as *Russula heterophylla*, *R. aeruginea*, and *R. virescens*, can be distinguished by their brittle flesh and the lack of both volva and ring. Other similar species include *A. subjunquillea* in eastern Asia and *A. arocheae*, which ranges from Andean Colombia north at least as far as central Mexico, both of which are also poisonous.

In January 2012, four people were accidentally poisoned when death caps (reportedly misidentified as straw fungi, which are popular in Chinese and other Asian dishes) were served at a New Year's Eve dinner party in Canberra, Australia. All the victims required hospital treatment and two of them died, with a third requiring a liver transplant.

Biochemistry

The species is now known to contain two main groups of toxins, both multicyclic (ring-shaped) peptides, spread throughout the mushroom tissue: the amatoxins and the phallotoxins. Another toxin is phallolysin, which has shown some hemolytic (red blood cell–destroying) activity *in vitro*. An unrelated compound, antamanide, has also been isolated.

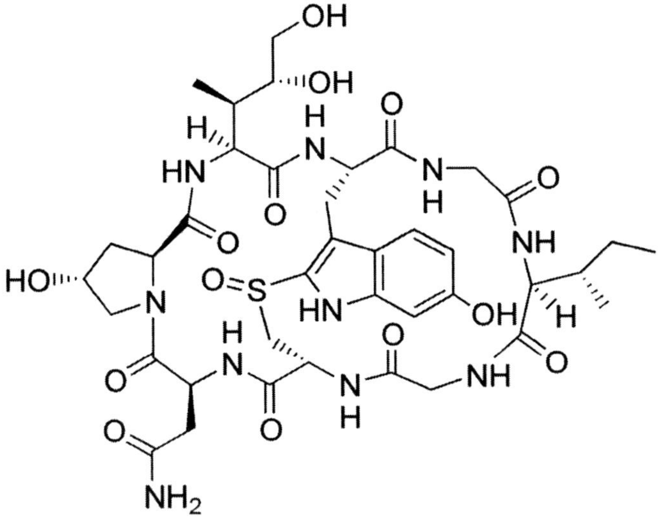

Figure: *α-amanitin*

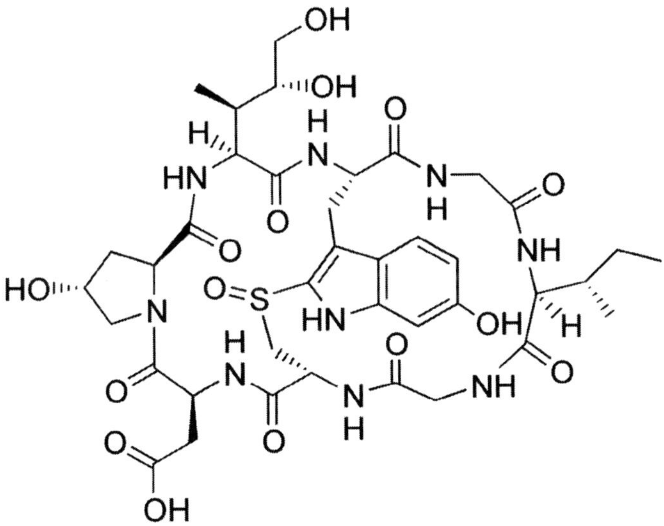

Figure: *β-Amanitin, where an amide of the á-amanitin has been replaced by a carboxylic acid*

Amatoxins consist of at least eight compounds with a similar structure, that of eight amino-acid rings; they were isolated in 1941 by Heinrich O. Wieland and Rudolf Hallermayer of the University of Munich. Of the amatoxins, á-amanitin is the chief component and along with â-amanitin is likely responsible for the toxic effects. Their major toxic mechanism is the inhibition of RNA polymerase II, a vital enzyme in the synthesis of messenger RNA (mRNA), microRNA, and small nuclear RNA (snRNA). Without mRNA, essential protein synthesis and hence cell metabolism grind to a halt and the cell dies. The liver is the principal organ affected, as it is the organ which is first encountered after absorption in the gastrointestinal tract, though other organs, especially the kidneys, are susceptible. The RNA polymerase of *Amanita phalloides* is insensitive to the effects of amatoxins, so the mushroom does not poison itself.

The phallotoxins consist of at least seven compounds, all of which have seven similar peptide rings. Phalloidin was isolated in 1937 by Feodor Lynen, Heinrich Wieland's student and son-in-law, and Ulrich Wieland of the University of Munich. Though phallotoxins are highly toxic to liver cells, they have since been found to add little to the death cap's toxicity, as they are not absorbed through the gut. Furthermore, phalloidin is also found in the edible (and sought-after) Blusher (*Amanita rubescens*). Another group of minor active peptides are the virotoxins, which consist of six similar monocyclic heptapeptides. Like the phallotoxins, they do not induce any acute toxicity after ingestion in humans.

Symptoms

Death caps have been reported to taste pleasant. This, coupled with the delay in the appearance of symptoms—during which time internal organs are being severely, sometimes irreparably, damaged—makes it particularly dangerous. Initially, symptoms are gastrointestinal in nature and include colicky abdominal pain, with watery diarrhea and vomiting, which may lead to dehydration, and, in severe cases, hypotension, tachycardia, hypoglycemia, and acid–base disturbances. These first symptoms resolve two to three days after the ingestion. A more serious deterioration signifying liver involvement may then occur—jaundice, diarrhea, delirium, seizures, and coma due to fulminant hepatic failure and attendant hepatic encephalopathy caused by the accumulation of normally liver-removed substance in the blood. Renal failure (either secondary to severe hepatitis or caused by direct toxic renal damage) and coagulopathy may appear during this stage. Life-threatening complications include increased intracranial pressure, intracranial hemorrhage, pancreatitis, acute renal failure, and cardiac arrest. Death generally occurs six to sixteen days after the poisoning.

Mushroom poisoning is more common in Europe than in America. Up to the mid-20th century, the mortality rate was around 60–70%, but this has been greatly reduced with advances in medical care. A review of death cap poisoning throughout Europe from 1971 to 1980 found the overall mortality rate to be 22.4% (51.3% in children under ten and 16.5% in those older than ten). This has fallen further in more recent surveys to around 10–15%.

Treatment

Figure: *Young "death cap" mushrooms in Poland, with matchbox for scale*

Consumption of the death cap is a medical emergency requiring hospitalization. The four main categories of therapy for poisoning are preliminary medical care, supportive measures, specific treatments, and liver transplantation.

Preliminary care consists of gastric decontamination with either activated carbon or gastric lavage. However, due to the delay between ingestion and the first symptoms of poisoning, it is commonplace for patients to arrive for treatment many hours after ingestion, potentially reducing the efficacy of these interventions. Supportive measures are directed towards treating the dehydration which results from fluid loss during the gastrointestinal phase of intoxication and correction of metabolic acidosis, hypoglycemia, electrolyte imbalances, and impaired coagulation.

No definitive antidote is available, but some specific treatments have been shown to improve survivability. High-dose continuous intravenous penicillin G has been reported to be of benefit, though the exact mechanism is unknown, and trials with cephalosporins show promise. Some evidence indicates intravenous silibinin, an extract from the blessed milk thistle (*Silybum marianum*), may be beneficial in reducing the effects of death cap poisoning. A long-term clinical trial of intravenous silibinin has just begun in the US. Silibinin prevents the uptake of amatoxins by hepatocytes, thereby protecting undamaged hepatic tissue; it also stimulates DNA-dependent RNA polymerases, leading to an increase in RNA synthesis. SLCO1B3 has been identified as the human hepatic uptake transporter for amatoxins; moreover, substrates and inhibitors of that protein—among others rifampicin, penicillin, silibinin, antamanide, paclitaxel, ciclosporin and prednisolone—may be useful for the treatment of human amatoxin poisoning.

N-Acetylcysteine has shown promise in combination with other therapies. Animal studies indicate the amatoxins deplete hepatic glutathione; N-acetylcysteine serves as a glutathione precursor and may therefore prevent reduced glutathione levels and subsequent liver damage. None of the antidotes used have undergone prospective, randomized clinical trials, and only anecdotal support is available. Silibinin and N-acetylcysteine appear to be the therapies with the most potential benefit.

Repeated doses of activated carbon may be helpful by absorbing any toxins returned to the gastrointestinal tract following enterohepatic circulation. Other methods of enhancing the elimination of the toxins have been trialed; techniques such as hemodialysis, hemoperfusion, plasmapheresis, and peritoneal dialysis have occasionally yielded success, but overall do not appear to improve outcome.

In patients developing liver failure, a liver transplant is often the only option to prevent death. Liver transplants have become a well-established option in amatoxin poisoning. This is a complicated issue, however, as transplants themselves may have significant complications and mortality; patients require long-term immunosuppression to maintain the transplant. That being the case, the criteria have been reassessed, such as onset of symptoms, prothrombin time (PTT), serum bilirubin, and presence of encephalopathy, for determining at what point a transplant becomes necessary for survival. Evidence suggests, although survival rates have improved with modern medical treatment, in patients with moderate to severe poisoning, up to half of those who did recover suffered permanent liver damage. However, a follow-up study has shown most survivors recover completely without any sequelae if treated within 36 hours of mushroom ingestion.

Notable Victims

Several historical figures may have died from *A. phalloides* poisoning (or other similar, toxic *Amanita* species). These were either accidental poisonings or assassination plots. Alleged victims of this kind of poisoning include Roman Emperor Claudius, Pope Clement VII, Tsaritsa Natalia Naryshkina, and Holy Roman Emperor Charles VI.

R. Gordon Wasson recounted the details of these deaths, noting the likelihood of *Amanita* poisoning. In the case of Clement VII, the illness that led to his death lasted some five months, making the case clearly inconsistent with amatoxin poisoning. Natalia Naryshkina is said to have consumed a large quantity of pickled mushrooms prior to her death. However, it is unclear whether the mushrooms themselves were poisonous or she succumbed to food poisoning.

Charles VI experienced indigestion after eating a dish of sautéed mushrooms. This led to an illness from which he died 10 days later—symptomatology consistent with amatoxin poisoning. Charles' death led to the War of the Austrian Succession. Noted Voltaire, "this dish of mushrooms changed the destiny of Europe."

The case of Claudius' poisoning is more complex. Claudius was known to have been very fond of eating Caesar's mushroom. Following his death, many sources have attributed it to his being fed a meal of death caps instead of Caesar's mushrooms. However, ancient authors, such as Tacitus and Suetonius, are unanimous about poison having been added to the mushroom dish, rather than the dish having been prepared from poisonous mushrooms. Wasson speculated the poison used to kill Claudius was derived from death caps, with a fatal dose of colocynth being administered later during his illness.

Amanita Porphyria

Amanita porphyria, also known as the Grey Veiled Amanita, is a fairly common, inedible basidiomycete mushroom of the genus *Amanita* found in Europe and North America.

Description

The cap is hemispherical when young, later flat. It is 4–10 cm in diameter, greyish brown to brown in colour, usually smooth, sometimes with grey patches. The gills are white and free from the stem. The spores are white. The stem is 5–12 cm high and 1-1.5 cm thick and has a grey to black ring. The stem has a basal bulb. The flesh is white with a smell of radish. *A. porphyria* usually grows near coniferous trees.

Uses

A. porphyria is not suitable for consumption. It is considered slightly poisonous and, more importantly, can easily be confused with much more poisonous species such as the Panther cap *(Amanita pantherina).*

Amanita Regalis

Amanita regalis, commonly known as the royal fly agaric or the king of Sweden Amanita, is a species of fungus in the Amanitaceae family. Common in Scandinavian countries, it is also found in eastern and northern Europe. In North America, its distribution is restricted to Alaska. The fruit bodies of the fungus somewhat resemble the fly agaric (*Amanita muscaria*), and it was formerly regarded as a variety of this species. *A. regalis* differs from it in being larger, with a liver-brown cap bearing numerous scabs, and in having a stem which is yellow-ochre at the base, with patches or rings of patches. Chemical analysis has shown that this species contains muscimol, the same psychoactive compound found in *A. muscaria*.

Taxonomy and Naming

Amanita regalis was first described as *Agaricus muscarius* â *regalis* by Elias Magnus Fries in his *Systema Mycologicum*, published in 1821. In 1887, Pier Andrea Saccardo treated it as a variety of *Amanita muscaria*. Edmund Michael, in 1903, became the first to consider it a distinct species. In 1941, Jean-Edouard Gilbert suggested a complete reorganization of the genus *Amanita* in his world monograph of the genus, and moved it to *Amanitaria* as *A. muscaria* var. *regalis*. In his original (1949) version of *Agaricales in Modern Taxonomy*, Rolf Singer considered it a subspecies of *A. muscaria*, but noted that it may be regarded as a separate species; in the fourth edition (1986), he listed it as a distinct species. *A. regalis* is classified in a section of *Amanita* within the genus, a grouping of related Amanitas that have a ring on the stem (or remnants thereof), and a bulb at the base of it. More recently, a Japanese group studied the biogeography of *A. muscaria* and related species, and, using molecular phylogenetic analysis, concluded that the taxon should be considered a grouping of *A. muscaria*, rather than a distinct species. However, as of 2012, both Index Fungorum and MycoBank list the taxon as *Amanita regalis*.

The species has been called by several vernacular names, including the "royal fly agaric", the "king of Sweden Amanita", the "brown fly agaric", or the king fly agaric. In France it is known as *Amanite royale*, while in Germany it is *Königsfliegenpilz*. The specific epithet is derived from the Latin word *regalis*, meaning "royal". In 2000, it was selected by the German Mycological Society as "Mushroom of the Year".

Description

Like all *Amanita* species, the bulk of the organism lies beneath the ground as a symbiotic partner to certain species of trees. The fruit body of the fungus is a reproductive structure that appears when appropriate environmental conditions of moisture, temperature, and nutrient availability are met. The cap of *A. regalis* is 10 to 25 cm (3.9 to 9.8 in) broad, and, depending on the stage of its development, can range in shape from spherical to convex to somewhat flattened. The color is yellowish brown, and it is densely covered with yellowish to light ochraceous scabby warts, which are arranged in almost regular concentric rings. The rings are a remnant of the volva left behind during expansion of the young fruit body. The cap is fleshy and when mature has grooves on the margin, which can extend 1.5 to 2 cm (0.6 to 0.8 in). The cap cuticle may be separated from the cap by peeling, to almost the center.

The gills are crowded close together, free from attachment to the stem, and white with a creamy yellow tinge. The edges of the gills are floccose, meaning they have tufts of soft wooly hairs—another volval remnant. The stem of a mature individual is typically between 10 and 14 cm (3.9 and 5.5 in) long and 1.5 to 2 cm (0.6 to 0.8 in) wide, and spreads at the base into a bulb ornamented with 2–4 rings of small squamulose, lemon or ochre-yellow warts. The entire stem and ring are not pure white but have a faint yellowish tinge. The flesh is whitish, faintly yellowish in the stem, and golden yellow under the cuticle of the cap. It does not change color in air, and has an insignificant taste and smell.

Microscopic Features

The spores are broadly ellipsoid to roughly spherical, hyaline (translucent), smooth, and have dimensions of 9–12 by 7–8 μm. They are non-amyloid, meaning that they do not absorb iodine when stained with Melzer's reagent. The spore-producing cells, the basidia, are club-shaped, measure 38–46 by 3–13 μm, and have clamps at their bases.

Similar Species

Amanita regalis is easily distinguished from *A. muscaria* by the absence of any red color in the cap, and the yellow patches on the stem. If it is a form that has a pale-colored, yellowish-brown cap, *Amanita regalis* may be confused with the blusher (*Amanita rubescens*, edible). The latter species can be identified by the flesh-colored tinge it stains when cut or bruised, and also by the color of the flesh underneath the cuticle: *A. regalis* is yellowish while *A. rubescens* is white. *Amanita pantherina* has a similar color, but has white flesh below the cap cuticle, and a cup-shaped bulb with distinct margins around the edge.

Habitat and Distribution

Amanita regalis is a rare species typically found growing on the ground in mountainous forests, both deciduous and coniferous. It is a mycorrhizal fungus, and exists in a symbiotic relationship with certain tree species. The system of fine below-ground threads, the mycelium, envelopes the roots of the trees and supplies them with needed minerals, trace elements and water from the soil, while the tree, in turn, supplies the fungus with food from its own supply, which it produces through photosynthesis. *A. regalis* has been shown experimentally to form mycorrhiza with birch, Scots pine, Mountain pine, and Norway spruce. Fruit bodies have been reported to grow in fairy rings.

In Europe it is more common in northern than south Europe, and it has not been reported from south and west Europe. In addition to being common in Scandinavian countries, it has been collected in Germany, Hungary, Latvia, Russia, Slovakia, and Korea. In North America, its distribution is restricted to Alaska, where it is usually found above the tree line.

Toxicity

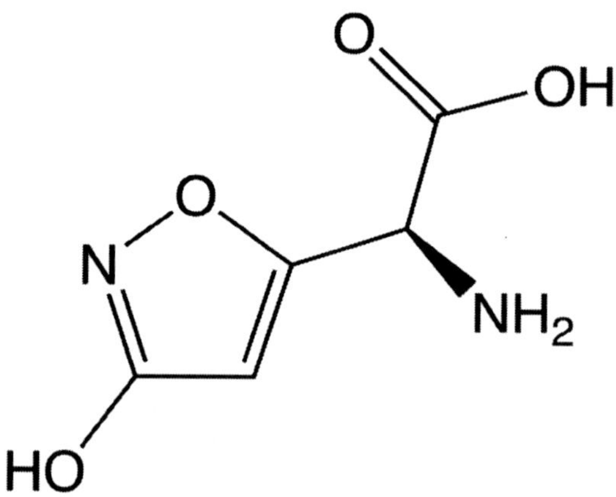

Figure: *Ibotenic acid*

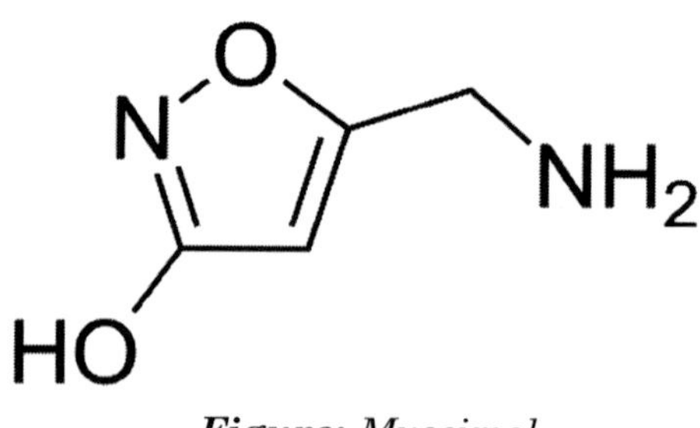

Figure: *Muscimol*

Amanita regalis is poisonous. A case of poisoning was reported from Finland, where three individuals believed they had consumed the edible parasol mushroom *Macrolepiota procera*. The symptoms of poisoning, which began 1–2 hours after ingestion of the mushrooms, were gastrointestinal—nausea and heavy vomiting. Two had central nervous system manifestations and cholinergic symptoms, including hallucinations, confusion, or loss of consciousness as well as profuse salivation and sweating. All three individuals recovered within 4–24 hours without any damage to liver, kidneys or the central nervous system. As this incident demonstrates, cooking the mushrooms does not completely neutralize the toxic components of *Amanita regalis*. Chemical analysis has shown that this species contains ibotenic acid and muscimol, the same toxic constituents as *Amanita muscaria*.

Amanita regalis has the ability to bioaccumulate the heavy metal vanadium, a phenomenon first reported in *Amanita muscaria* in 1931. A field study of Scandinavian specimens found vanadium contents ranging from 38 to 169 mg of vanadium per kg of dry mushroom (average of 119 mg/kg). For comparison, the vanadium concentration in most other mushrooms is typically less than 2 mg/kg.

Amanita Smithiana

Amanita smithiana, also known as Smith's Amanita, is a species of agaric found on soil in coniferous (*Abies*, *Tsuga*, *Pseudotsuga*) and broadleaved (*Alnus*, *Quercus*) woodland in the Pacific Northwest of North America. It fruits in August and September. The cap has a diameter of 5–17 cm (2–7 in) and is white and scaled with remnants of the universal veil. The stem is 6–18 cm (2–7 in) and white and similarly scaled, with a ring. It is responsible for poisonings in the Pacific Northwest when mistaken for the edible and sought after *Tricholoma magnivelare*. It causes initial gastrointestinal symptoms, followed by acute renal failure after a delay of 2-6 days. This is often severe, requiring hemodialysis, but most patients recover normal kidney function within several weeks.

Amanita smithiana was described by Dutch mycologist Cornelis Bas in 1969. It belongs in the subgenus *Lepidella*. Two similar species have been implicated in similar cases of acute renal failure; *A. proxima* in Spain and *A. pseudoporphyria* in Japan.

Amanita Subjunquillea

Amanita subjunquillea, also known as the East Asian death cap is a mushroom of the large genus *Amanita*, which occurs in East and

Southeast Asia. Deadly poisonous, it is a member of section phalloideae and related to the death cap *A. phalloides*.

Initially little reported, the toxicity of *A. subjunquillea* has been well established; a study in Korea revealed it to have similar effects to *A. phalloides*, namely delayed gastrointestinal symptoms, hepatotoxicity and a 12.5% mortality.

An all-white variety, *Amanita subjunquillea* var. *alba* is known from southwestern China, Japan and Northern India.

Amanita Verna

Amanita verna, commonly known as the fool's mushroom, Destroying angel or the mushroom fool, is a deadly poisonous basidiomycete fungus, one of many in the genus *Amanita*. Occurring in Europe in spring, *A. verna* associates with various deciduous and coniferous trees. The large fruiting bodies (*i.e.*, the mushrooms) appear in summer and autumn; the caps, stipes and gills are all white in colour.

Initially described by the French botanist Jean Baptiste François Pierre Bulliard, the fool's mushroom's specific epithet *verna* is derived from its springtime fruiting habit.

Description

The fool's mushroom (*Amanita verna*) also known as the spring destroying angel or death angel, is a close relative of Amanita phalloides, the death cap and a member of the mushroom genus *Amanita*. *Amanita verna*, like its close relative, belongs to the subfamily Phalloideae.

The fool's mushroom is pure white, all the way to the gills and the stem. This fungus, like all amanitas, has a volva. The fool's mushroom's cap is 5–10 centimetres (2–4 in) wide, and is about the same height. This mushroom's lamellae is free and white, and the volva is bag-like and large. Its annulus is white and membranous, and *A. verna* also does not react with potassium hydroxide solution, unlike its relative *Amanita virosa*. The mushroom's spores are smooth and elliptical.

Habitat and Season

The mushroom lives in Europe. The fool's mushroom is known to grow in woodlands and hardwood forests. As the fool's mushroom other name (spring destroying angel) suggests, this toxic amanita grows in spring. Unlike various closely related poisonous amanitas, this mushroom is not known to occur in North America.

Toxicity

Closely related to other deadly pure white amanitas, the fool's mushroom is one of the most poisonous mushrooms in the world. Just like the death cap, it contains amatoxins, primarily alpha-amanitin, which can cause liver failure. While this mushroom also contains other poisons like phallotoxins, these toxins are not behind the fatal poisoning that this mushroom, as well as all members of the subfamily Phalloideae, causes.

This mushroom's toxicity and symptoms are similar, if not identical to that of the death cap. Like other members of the subfamily Phalloideae, the fool's mushroom has been implicated in a number of serious or fatal poisonings.

There are no negative symptoms from eating this fungus until 6–24 hours after ingestion. The first symptom is simply unease. Violent cramps and diarrhea follow. On the third day, the same symptoms repeat themselves, but while to many this may seem like a sign of recovery, most of the time it is simply a herald of the final onset of symptoms, which include kidney and liver failure due to amatoxins. At this point, drastic measures like liver transplants need to be taken, or the victim would most likely die.

The fatal dosage of this mushroom is around 30g (1 oz), though it is not recommended taking even the smallest amount of any deadly amanita, as the fatal dosage varies with the size of the person who consumes the mushroom.

Similar Species

Edible Species: Many edible species resembles this mushroom and other amanitas, which is why the consumption of amanitas accounts for 95% of all mushroom fatalities.

The edible mushroom *Leucoagaricus leucothites* very closely resembles the fool's mushroom. However, *Leucoagaricus leucothites* lacks a volva, unlike *A. verna*. Sometimes, the volva of *A. verna* is hidden under leaf litter or soil, raising the need to dig out rather than cut *L.naucina* from the ground, so as to prevent misidentification.

Various young edible *Agaricus spp.* closely resemble *A. verna*, such as the tasty field mushroom *Agaricus campestris*, *Agaricus silvicola* and *Agaricus arvensis*. While older specimens of these do not resemble *A. verna*, when these fungus are young, they do resemble *A. verna*. All of these mushrooms lack volvas, unlike their deadly lookalike.

Amanita caesarea, with its bright orange shade, does not resemble *A. verna*, while this fungus is in bulb stage, it is also pure white. Amanitas closely resemble each other in bulb stage.

Another group of mushrooms in the *Volvariella* genus are frequently mistaken for the deadly amanitas. *Volvariella volvacea* has often been mistaken for deadly amanitas, and most of the time, they are mistaken for a relative of the fool's mushroom, the death cap (*Amanita phalloides*). *Volvariella speciosa* looks very similar to the fool's mushroom, but with close examination, one would realise that *V.speciosa* has salmon gills, while the fool's mushroom has white gills. Like *Amanita*, *Volvariella* species have volvas, hence their name.

Amanitopsis vaginata, formerly *Amanita vaginata*, can be pure white, and thus sometimes looks nearly identical to the fool's mushroom. However, unlike all *Amanita* species, *Amanitopsis* species lack rings. However, the ring of *A. verna* may sometimes be eaten by worms, and thus may be missing. While in bulb stage, *A. verna*, as well as its relatives, are commonly mistaken for the edible puffballs, like those form the *Lycoperdon* or *Calvatia* species. To prevent confusion, it is important to cut puffballs into half. If there is the outline of *Amanita*'s cap and stem in the 'puffball', the mushroom is an amanita.

The fatal poisoning caused by the fool's mushroom and closely related species makes amateur mushroom hunters avoid any species similar to them, and buy them form a grocer instead. Unless one can truly be sure if the species he or she is digging is edible, one should never pick a similar mushroom.

Poisonous Species

Of course, the other pure white amanitas, like the destroying angels (*Amanita virosa, Amanita bisporigera, Amanita ocreata*) closely resemble *Amanita verna*. However, unlike *Amanita verna, Amanita virosa* stains yellow in KOH. Pure white specimens of the death cap also look like the fool's mushroom, and sometimes, *Amanita verna* is considered a subspecies of *Amanita phalloides*.

Other toxic species of *Agaricus*, such as *Agaricus xanthodermus*, also resemble *A. verna*. However, *Agaricus xanthodermus* stains yellow if bruised, gives off a repulsive smell when cooked, and most importantly, lacks a volva.

List of Similar Species

Agaricus arvensis (Horse Mushroom) : excellent edible

Agaricus campestris (Field mushroom) : excellent edible

Agaricus silvicola (Wood mushroom) : excellent edible

Agaricus xanthodermus (Yellow-staining mushroom) : toxic or at least unpleasant-tasting (varies with consumer)

Amanita bisporigera (Destroying angel) : deadly poisonous

Amanita caesarea (Caesar's Amanita) (bulb stage only) : excellent edible

Amanita ocreata (American destroying angel) : deadly poisonous

Amanita phalloides (death cap) : deadly poisonous

Amanita virosa (European destroying angel) : deadly poisonous

Amanitopsis vaginata (Grisette) : Edible

Calvatia spp. (Giant puffballs) : (bulb stage only) : Mostly edible when young

Leucoagaricus leucothites : Edible and good, but can cause mild poisonings in some individuals

Lycoperdon spp. (Puffballs) : Mostly edible when young

Volvariella speciosa : Edible but not tasty

Volvariella volvacea (Paddy straw mushroom) : Edible and good

Amanita Virosa

Amanita virosa, commonly known as the European destroying angel, is a deadly poisonous basidiomycete fungus, one of many in the genus *Amanita*. Occurring in Europe, *A. virosa* associates with various deciduous and coniferous trees. The large fruiting bodies (*i.e.*, the mushrooms) appear in summer and autumn; the caps, stipes and gills are all white in colour.

Immature specimens of *A. virosa* resemble several edible species commonly consumed by humans, increasing the risk of accidental poisoning. Along with its geographical namesakes, *A. virosa* is one of the most poisonous of all known poisonous mushrooms; its principal toxic constituent á-amanitin damages the liver and kidneys, often fatally. There are only a few known antidotes, one of which is a compound derived from milk thistle.

Taxonomy and Naming

The common name of *destroying angel* is applied to several all-white species of poisonous *Amanita*, to this species in Europe and to *Amanita bisporiga* in eastern North America, and *A. ocreata* in the west. *A. virosa* was first collected and described by Elias Magnus Fries in Sweden. Its specific epithet *virosa* derived from the Latin adjective *virôsus* 'toxic' (compare virus).

Amanita virosa is very similar to several other species of all-white amanitas known as destroying angels, which has led to confusion over which occurs where. This specific name has been applied to all-white destroying angels occurring in North America, though others propose these all belong to *A. bisporiga* and other rarer species instead. There has been some question over whether *Amanita verna* is a valid species.

Description

A. virosa first appears as a white egg-shaped object covered with a universal veil. As it grows, the mushroom breaks free, though there may be ragged patches of veil at the cap edges. The cap is initially conical with inturned edges, before becoming hemispherical and flattening with a diameter up to 12 cm (4½ in). The cap often has a distinctive boss; it is able to be peeled and white, though the centre may be ivory in colour. The crowded free gills are white, as is the stipe and volva. The thin stipe is up to 15 cm (6 in) tall, with a hanging grooved ring. The spore print is white and the spores egg-shaped conical and 7–10 ìm long. They stain blue with iodine. The flesh is white, with a taste reminiscent of radishes, and turns bright yellow with sodium hydroxide.

This fungus highlights the danger of picking immature fungi as it resembles the edible mushrooms *Agaricus arvensis* and *A. campestris*, and the puffballs (*Lycoperdon* spp.) before the caps have opened and the gills have become visible. The ability to be peeled has been taken as a sign of edibility in mushrooming, which is a potentially lethal mistake in this species. It is unclear why this fungus which more closely resembles edible species has been implicated in fewer deaths than the death cap, though its rarity may contribute to this.

Distribution and Habitat

A. virosa is found in mixed woodland, especially in association with beech, on mossy ground in summer and autumn. Most *Amanita* species form ectomycorrhizal relationships with the roots of certain trees.

Toxicity

Amanita virosa is highly toxic, and has been responsible for severe mushroom poisonings. Like the closely related death cap (*A. phalloides*), it contains the highly toxic amatoxins, as well as phallotoxins. Some authorities strongly advise against putting these fungi in the same basket with those collected for the table and to avoid touching them.

Figure: *young fruiting bodies showing conical caps*

Amatoxins consist of at least eight compounds with a similar structure, that of eight amino-acid rings; they were isolated in 1941 by Heinrich O. Wieland and Rudolf Hallermayer of the University of Munich. Of the amatoxins, á-amanitin is the chief component and along with â-amanitin is likely responsible for the toxic effects. Their major toxic mechanism is the inhibition of RNA polymerase II, a vital enzyme in the synthesis of messenger RNA (mRNA), microRNA, and small nuclear RNA (snRNA). Without mRNA essential protein synthesis and hence cell metabolism grind to a halt and the cell dies. The liver is the principal organ affected, as it is the organ which is first encountered after absorption in the gastrointestinal tract, though other organs, especially the kidneys, are susceptible.

The phallotoxins consist of at least seven compounds, all of which have seven similar peptide rings. Phalloidin was isolated in 1937 by Feodor Lynen, Heinrich Wieland's student and son-in-law, and Ulrich Wieland of the University of Munich. Though phallotoxins are highly toxic to liver cells, they have since been found to have little input into the destroying angel's toxicity as they are not absorbed through the gut. Furthermore, phalloidin is also found in the edible (and sought-after) Blusher (*Amanita rubescens*). Another group of minor active peptides are the virotoxins, which consist of six similar monocyclic heptapeptides. Like the phallotoxins they do not exert any acute toxicity after ingestion in humans.

Treatment

Consumption of *Amanita virosa* is a medical emergency requiring hospitalization. There are four main categories of therapy for poisoning: preliminary medical care, supportive measures, specific treatments, and liver transplantation.

Preliminary care consists of gastric decontamination with either activated carbon or gastric lavage. However, due to the delay between ingestion and the first symptoms of poisoning, it is commonplace for patients to arrive for treatment many hours after ingestion, potentially reducing the efficacy of these interventions. Supportive measures are directed towards treating the dehydration which results from fluid loss during the gastrointestinal phase of intoxication and correction of metabolic acidosis, hypoglycemia, electrolyte imbalances, and impaired coagulation.

No definitive antidote for amatoxin poisoning is available, but some specific treatments have been shown to improve survivability. High-dose continuous intravenous penicillin G has been reported to be of benefit, though the exact mechanism is unknown, and trials with cephalosporins show promise. There is some evidence that intravenous silibinin, an extract from the blessed milk thistle (*Silybum marianum*), may be beneficial in reducing the effects of death cap poisoning. Silibinin prevents the uptake of amatoxins by hepatocytes, thereby protecting undamaged hepatic tissue; it also stimulates DNA-dependent RNA polymerases, leading to an increase in RNA synthesis. N-acetylcysteine has shown promise in combination with other therapies. Animal studies indicate the amatoxins deplete hepatic glutathione; N-acetylcysteine serves as a glutathione precursor and may therefore prevent reduced glutathione levels and subsequent liver damage. None of the antidotes used have undergone prospective, randomized clinical trials, and only anecdotal support is available. Silibinin and N-acetylcysteine appear to be the therapies with the most potential benefit. Repeated doses of activated carbon may be helpful by absorbing any toxins that are returned to the gastrointestinal tract following enterohepatic circulation. Other methods of enhancing the elimination of the toxins have been trialed; techniques such as hemodialysis, hemoperfusion, plasmapheresis, and peritoneal dialysis have occasionally yielded success but overall do not appear to improve outcome.

In patients developing liver failure, a liver transplant is often the only option to prevent death. Liver transplants have become a well-established option in amatoxin poisoning. This is a complicated issue, however, as transplants themselves may have significant complications and mortality; patients require long-term immunosuppression to maintain the transplant. That being the case, there has been a reassessment of criteria such as onset of symptoms, prothrombin time (PTT), serum bilirubin, and presence of encephalopathy for determining at what point a transplant becomes necessary for survival. Evidence suggests that, although survival rates have improved with modern medical treatment, in patients with moderate to severe poisoning up to half of those who did recover suffered permanent liver damage. However, a follow-up study has shown that most survivors recover completely without any sequelae if treated within 36 hours of mushroom ingestion.

Amanita Xanthocephala

The vermilion grisette, also known as pretty grisette or vermilion *Amanita* (*Amanita xanthocephala*) is a colourful mushroom of the genus *Amanita*. However, although it is often referred to by the common name "grisette", it is not closely related to other edible species that carry this common name, such as *Amanita vaginata* and *Amanita fulva*. It belongs to the same group of *Amanita* as *A. muscaria* and is reported to be toxic.

It is a ringless mushroom with a yellowish- to reddish-orange cap up to 3–5 cm (1–2 in) in diameter, with deeper colour toward the centre, and paler similar-coloured warts. The gills and slim ringless stipe are pale yellow or white. The white volva has a neat outturned lip and is often bordered with orange or yellow.

Amanita xanthocephala lives in an ectomycorrhizal relationship with *Eucalyptus*. It is found in eucalypt forests in the southwest of Western Australia, as well as southeastern Australia from near Adelaide around to Southeast Queensland.

At one stage this fungus was known as *Amanitopsis pulchella*, a small genus that all grisettes (ringless *Amanita* species) were placed in. However, this genus has been later sunk back into *Amanita*. Unlike most ringless *Amanita*, which are part of *Amanita* section *Vaginatae* (*A. vaginata* and allies), *A. xanthcephala* belongs to *Amanita* section *Amanita* (*A. muscaria* and allies).

There is one report of a person being quite ill after tasting a small piece of it in 1997.

Destroying Angel

The name destroying angel applies to several similar, closely related species of deadly all-white mushrooms in the genus *Amanita*. They are *Amanita bisporigera* and *A. ocreata* in eastern and western North America, and *A. virosa* in Europe. Another very similar species, *A. verna* or fool's mushroom was first described in France.

Closely related to the death cap (*A. phalloides*) they are among the most toxic known mushrooms, containing amatoxins as death caps do.

Description

***Figure:** A half-grown destroying angel*

Destroying angels are characterized by having a white stalk and gills. The cap can be pure white, or white at the edge and yellowish, pinkish, or tan at the center. It has a partial veil, or ring (annulus) circling the upper stalk, and the gills are "free," not attached to the stalk. Perhaps the most telltale of the features is the presence of a volva, or universal veil, so called because it is a membrane that encapsulates the entire mushroom, rather like an egg, when it is very young. This structure breaks as the young mushroom expands, leaving parts that can be found at the base of the stalk as a boot or cuplike structure, and there may be patches of removable material on the cap surface. This combination of features, all found together in the same mushroom, is the hallmark of the family. While other families may have any one or two of these features, none has them all. The cap is usually about 5–12 cm across; the stipe is usually 7½–20 cm long and about ½–2 cm thick. They are found singly or in small groups.

Destroying angels can be mistaken for edible fungi such as the button mushroom, meadow mushroom, or the horse mushroom. Young destroying angels that are still enclosed in their universal veil can be mistaken for puffballs, but slicing them in half longitudinally will reveal internal mushroom structures. This is the basis for the common recommendation to slice in half all puffball-like mushrooms picked when mushroom hunting. Mushroom hunters recommend that people know how to recognize both the death cap and the destroying angel in all of their forms before collecting any white gilled mushroom for consumption.

Distribution and Habitat

All *Amanita* species form ectomycorrhizal relationships with the roots of certain trees. Thus destroying angels grow in or near the edges of woodlands. They can also be found on lawns or grassy meadows near trees or shrubs. Several species called Destroying Angel are found all over the world.

Toxicity

The destroying angel (*Amanita bisporigera*) and the death cap (*Amanita phalloides*) are responsible for the overwhelming majority of deaths due to mushroom poisoning. The toxin responsible for this is amatoxin. Symptoms do not appear for 5 to 24 hours, when the toxins may already be absorbed and the damage (destruction of liver and kidney tissues) done. As little as half a mushroom cap can be fatal if the victim is not treated quickly enough. The symptoms include vomiting, cramps, delirium, convulsions, and diarrhea. In one experiment, people who had ingested the toxin were treated with "fluid and electrolyte replacement, oral activated charcoal and lactulose, IV penicillin, combined hemodialysis and hemoperfusion in two 8-hour sessions", some with "IV thioctic acid, others IV silibinin" and all received a "special diet." It was concluded that "...intensive combined treatment applied in these cases is effective in relieving patients with both moderate and severe amanitin poisoning."

Physicians have had success in treatment of amatoxin poisoning using "anti-hepatotoxic" compounds from the milk thistle, *Silybum marianum*. A crude extract of flavolignans from *S. marianum* seeds, called silymarin (trade name *Legalon*) has proven useful in amatoxin poisoning cases. In a trial of one of the flavolignans silybin, in 60 patients poisoned by amatoxin-containing *Amanita* species, there were no deaths. (see Der Marderosian & Liberti 1988 and Foster 1991 for a summary of this work).

Amatoxin binds to RNA polymerase II, one of enzymes responsible for expression of genes in human cells.

Harmful Algal Blooms

Harmful algal blooms commonly occur as surface scums and "blooms." Most occur in warm, still or slow moving, green-water lakes during summer and early fall months. They are often found in farm dams or ponds where very little mixing occurs, allowing warm water layers to form near the surface. Though many types of algae can form blooms, harmful algal blooms in lakes and other freshwaters have the ability to produce toxins that are dangerous to other organisms, including humans and aquatic life such as fish. Harmful algal blooms are formed by certain types of photosynthetic organisms called cyanobacteria or blue-green algae.

Harmful algal blooms can appear quickly and form ugly floating mats of various colors, or they can discolor the water. Blooms can also have a pungent, foul odor. While it is estimated most blooms are toxic, it is difficult to predict exactly when or even if a bloom is producing toxins. It takes a few days to sample blooms and test them for toxicity, and by that time the toxicity may have changed. Furthermore, the harmful effects of blooms may even occur when a surface scum is not visible. Blooms that occur in drinking water sources can produce compounds that lead to toxicity as well as taste and odor problems. Toxicity can even occur in drinking water without taste and odor compounds. Depending on the region, harmful blooms can be a significant concern if drinking water comes from reservoirs or natural lakes.

Some of the techniques used to control or remove blooms can increase the water's toxicity, such as the use of algaecides. They may also be detrimental to organisms other than cyanobacteria and are generally not recommended. The most effective method to reduce blooms is through broad community effort to reduce nutrient loading into lakes. This is most often achieved through reducing the input of phosphorus, as well as nitrogen in a more limited extent. Though cyanobacteria are found in some of the oldest fossils and are natural to aquatic habitats, harmful algal blooms are increasing in frequency in the United States and across the world and are becoming more of a problem as eutrophication increases.

Health Concerns

The toxins produced by harmful algal blooms are a concern for human health. Toxins are produced within the cells and remain there until the cells break open, which may lead to the unpredictable toxicity of blooms. Water treatment processes can sometimes remove cyanobacteria and the toxins they produce although sometimes further steps are necessary because certain toxins are more difficult to remove than others. Toxins can also break down naturally in lakes. The World Health Organization (WHO) says that 100,000 cells/mL is a moderate human health risk, but there are currently no standards for cell or toxin concentrations in the United States. The most common toxin is called microcystins. There are at least 60 different types of microcystins, and their toxicity can vary significantly.

The WHO standard for microcystin-LR in drinking water is 1.0μg/L, but there are no similar standards for waters in the United States. Other toxins found in harmful algal blooms include nodularin, anatoxins, lyngbya, and cylindrospermopsis. Nodularin are similar to microcystins and have the same effects. Anatoxins are usually more toxic than microcystins, and saxitoxins, the cause of paralytic shellfish poisons, are highly poisonous. Lyngbya toxin a and aplysiatoxins are both carcinogenic, and lyngbya toxin a causes seaweed dermatitis. Cylindrospermopsis is most harmful to the liver, but other organs may also be damaged by it.

Toxin Types	***Examples***	***Effects***
Neurotoxins	Anatoxin-a, anatoxin-a(s), saxitoxin, neosaxitoxin	Affects central nervous system, causes seizures, paralysis, respiratory failure, and death
Hepatotoxins	Microcystins, nodularins, cylindrospermopsin	Affects liver, causes hemorrhaging, tissue damage, tumors, liver cancer, and death
Dermatotoxins and Gastrointestinal toxins	Aplysiatoxins, lyngbyatoxin-a, lipopolysaccharide endotoxins	Affects skin and mucous membranes, causes rashes, respiratory illness, headache, and stomach upset
Cytotoxins	Cylindrospermopsin	Affects liver and other organs; causes chromosome loss, DNA strand breakage, and organ damage

Exposure

Exposure to cyanobacteria and cyanobacterial toxins can occur from the following:

- Drinking water that contains harmful algal bloom cells
- Drinking untreated water
- Engaging in recreational activities such as swimming in water experiencing a bloom
- Inhaling aerosolized cyanobacteria or toxins such as the spray from jet skiing and boating

- Using contaminated water to water lawns or irrigate golf courses
- Consuming cyanobacterial supplements containing mycrocystins
- Dialysis (this has only occurred in Brazil)

Effects of Harmful Algal Blooms

Decreased recreational use and aesthetic value of waters due to toxicity, mats of algae, and the smell when cells begin to die are only some of the problems associated with harmful algal blooms. Harmful algal blooms can contaminate drinking water with taste, odor, or toxic compounds. The toxins produced during blooms are possible carcinogens to humans, and current research is studying the link between certain cyanobacterial toxins and neurological diseases such as Lou Gehrig's disease. There has only been one human death in the US possibly related to recreational exposure to microcystins, though there are multiple cases of gastrointestinal distress, skin irritation and nausea. Recent reports have indicated that microcystins can be aerosolized, making recreational activities such as boating and jet-skiing a higher risk of exposure. Microcystins may act as tumor promoters and there is increasing evidence that these compounds may be more harmful through long term chronic exposure, such as low doses in drinking water, rather than short term acute exposure. This chronic toxicity is of significant concern to communities that are consuming drinking water with cyanotoxins in small doses over a lifetime.

Harmful algal blooms have been known to kill waterfowl and livestock, and dogs have died after eating mats of cyanobacteria or licking their fur after swimming in bloom-infested waters. In some cases, humans have also died after exposure to harmful algal blooms.

Harmful algal blooms negatively impact the food web by decreasing the amount of nutritious, edible phytoplankton that zooplankton and other primary consumers need to survive. These organisms may then starve, leading to decreased food for higher order consumers such as fish. Harmful algal blooms can block sunlight from phytoplankton under the water's surface, leading to decreased food and oxygen levels. When harmful algal blooms begin to break down and die, they can decrease dissolved oxygen levels, a change that can be lethal to other aquatic organisms and cause fish kills. Low dissolved oxygen can be made worse by overcast days and warmer temperatures.

Bibliography

Anthony B.: *Applied Ethnobotany: People, Wild Plant use, and Conservation*, London, Earthscan, 2001.

Arnberger. L. P. : *Flowers of the Southwest Mountains,* Southwestern Monuments Ass. 1968.

Bensky, D., Clavey, S., Stoger, E.: *Materia,* Medica Eastland Press, Inc. Seattle, 2004.

Bensky, D., Clavey, S., Stoger, E.: *Materia,* Medica Eastland Press, Inc. Seattle, 2004.

Chetan, A. and Brueton, D.: *The Sacred Yew*, Arkana, London, 1994.

Connor, H. E.: *Poisonous Plants in New Zealand,* Government Printer, Wellington, 1977.

Crowe, Andrew: *Which Native Forest Plant?*, Penguin, New Zealand, 1999.

Culpeper, Nicholas: *The English Physitian: or an Astrologo-physical Discourse of the Vulgar Herbs of this Nation*, Peter Cole, London, 1990.

Everist, Selwyn L.: *Poisonous Plants of Australia, Angus and Robertson*, Melbourne, 1974.

Everitt, J.H.; Lonard, R.L., Little, C.R.: *Weeds in South Texas and Northern Mexico*, Texas Tech University Press, Lubbock, 2007.

Funnell, K.A.: *Zantedeschia*, Elsevier Scientific Publishers, Amsterdam, 1993.

Giancarlo Pepeu, Maria Grazia Giovannini: *From Messengers To molecules: Memories are Made of these,* Springer, 2004.

Grunwald, Michael: *The Swamp*, Simon & Schuster, 2007.

Hartzell, H.: *The Yew Tree: A Thousand Whispers: Biography of a Species*, Hulogosi, Eugene, 1991.

Hussey, BMJ; Keighery GJ, Cousens RD, Dodd J, Lloyd SG: *Western Weeds: A Guide to the Weeds of Western Australia*, Plant Protection Society of WA, Perth, 1997.

Jones, David E.: *Poison Arrows: North American Indian Hunting and Warfare*, University of Texas Press, 2007.

Keddy, P.A. *Plants and Vegetation: Origins, Processes, Consequences*, Cambridge University Press, Cambridge, UK, 2007.

Lewis, W.H. and M.P.F. Elvin-Lewis: *Medical Botany. Plants Affecting Man's Health*, Wiley, New Yok, 1977.

McGee, Harold: *On Food and Cooking: The Science and Lore of the Kitchen*, New York, NY: Scribner, 2004.

Pittenger, Dennis R.: *California Master Gardener Handbook*, ANR Publications, 2002.

Pojar, Jim, A. MacKinnon, and Paul B. Alaback: *Plants of the Pacific Northwest Coast*, Lone Pine Pub., Redmond, WA, 1994.

Potter, Samuel O.L.: *A Handbook of Materia Medica Pharmacy and Therapeutics*, P. Blakiston's, London, 1893.

Robbins, W.W., M.K. Bellue and W.S. Ball: *Weeds of California*, State Department of Agriculture, Sacramento, California, 1941.

Spoerke, D.G. & Smolinske, S.C.: *Toxicity of Houseplants. CRC Press*, Inc., Boca Raton, Fla., USA, 1990.

Tilford, Gregory L.: *Edible and Medicinal Plants of the West*, Mountain Press Pub., Missoula, Montana, 1997.

Yeung. Him-Che. : *Handbook of Chinese Herbs and Formulas*, Institute of Chinese Medicine, Los Angeles 1985.

Index

G

H

I

J

K

M

N

P

R

S

T

V

W

❑❑❑